Fridtjof Nansen

The structure and combination of the histological elements of the central nervous system

Fridtjof Nansen

The structure and combination of the histological elements of the central nervous system

ISBN/EAN: 9783337276195

Printed in Europe, USA, Canada, Australia, Japan

Cover: Foto ©berggeist007 / pixelio.de

More available books at **www.hansebooks.com**

The Structure and Combination

of the

Histological Elements

of the Central Nervous System.

By

Fridtjof Nansen,
Curator of Bergen's Museum.

Reprinted from: Bergens Museums Aarsberetning for 1886.

BERGEN.
Printed by John Grieg.
1887.

The Structure and Combination

of the

Histological Elements
of the Central Nervous System.

By

Fridtjof Nansen,
Curator of Bergen's Museum.

Introduction.

I. History.

The progressive history of our knowledge of the histology of the nervous system is treated of so often, and so well, by previous writers that it certainly, to some extent, entails a repetition of their words to refer to it again.

Still, it is a tradition to look back upon the works and merits of our predecessors before passing to our own work, and I do not think it right, to break with the custom, but will, however, confine myself as much as possible to mentioning the most important writers of modern times on this subject, and for earlier literature on the subject refer my readers to the many sketches given by previous writers.[1] As it is especially the nervous system of the *evertebrates* which is about to be treated of here, I will restrict myself, principally, to the literature on that subject, all the more, as in a separate memoir on the nervous system of *Myxine glutinosa* I hope to obtain an opportunity of referring to the literature in respect of the histology of the *vertebrate* nervous system more circumstantially.

---

[1] The literature on the nervous system of the *evertebrates* is very circumstantially treated of by Vignal. His reports are, however, not in every point quite correct. Very good reviews of the literature are given 1882 by Freud, 1879 by Hans Schultze, 1875 by Hermann and 1872 by Solbrig. Of course, a great many other writers have also mentioned the previous literature more or less circumstantially. The previous literature on the histology of the nervous system of the *vertebrates* is referred to in Prof. Golgis last work on this subject. Hans Schultze has given a very good review of the literature so far as it concerns the fibrillous structure of the nervous elements. Besides this there is in Kuhnt's paper a very complete review of the literature up to his time.

a) The structure of the nerve-tubes.[1]

The structure of the nerve-tubes: whether it is fibrillous or non-fibrillous, has been very much disputed, and to this day the point must be considered as an open question. Since the valuable and important works of REMAK, there have, certainly, always been plenty of writers to defend the fibrillous structure. Amongst these may be named HANNOVER,[2] WILL, LEBERT and ROBIN, WALTER. WALDEYER, LEYDIG (only to a certain extent), BOLL, FLEMMING, HERMANN, DIETL, HANS SCHULTZE, FREUD, VIGNAL etc. Especially have HERMANN, HANS SCHULTZE and FREUD expressed themselves very emphatically and distinctly in favour of the fibrillous structure. According to their view the contents of the nerve-tubes consist in »Primitivfibrillen« suspended or swimming in an homogeneous semi-fluid »Interfibrillärsubstanz«.

At the same time, however, there have also been others who described and asserted a semi-fluid (fest-weiche) contents, and a homogeneous structure of the nerve-tubes in the fresh live state (or, also, a granulous structure). Amongst those are HELMHOLTZ,[3] HÆCKEL,[4] FAIVRE,[5] OWSJANNIKOW,[6] BUCHHOLZ, LEMOINE, SOL-

---

[1] I prefer the designation nerve-tube to that of nerve-fibre, because I think it more characteristic.

[2] Hannover states, that the nervous tubes of the *Mollusca* are longitudinally striated, whilst those of *Astacus* have a granular, nebulous substance in a membranous sheath.

[3] Helmholtz described the nervous tubes as »zarthäntige Cylinder mit flüssigen Inhalt«.

[4] According to Hæckel the nervous fibres (»Primitivröhre) have the form of tubes. »Die Wand der Primitivröhre oder die Nervenprimitivscheide ist an den feinern Cylindern einfach, an den stärkeren doppelt, immer aber scharf und dunkel contourirt, so dass sie sehr deutlich aus dem umhüllenden matten Bindegewebe hervorschimmert.« »Die Inhalt der Nervenprimitivröhre ist wie schon Helmholtz erkannte eine dicke Flüssigkeit, welche .... frish .... vollkommen homogen, wasserklar, und leicht glänzend erscheint.« Hæckel has, however, also found Remak's »centrale Faserbündel« and agrees with him, that it might be possible that such a one occurs in all nervous tubes.

[5] According to Faivre the contents of the nervous tubes are granulous. »Une tube de Sangsue se compose de deux parties: l'enveloppe et le contenu. L'enveloppe est anhiste, sans structure appréciable, sans noyaux, — le contenu des tubes est formé par une substance finement granuleuse et d'une consistance molle, meme à l'état frais.« Under high powers of magnification this substance is seen to consist of very minute granules »agglutinés par une sorte de substance intermediaire amorphe«.

[6] Owsjannikow supposes the fibrillous appearance of the tubes to be a postmortem product, and he does not believe, that »les tuyaux« described by Stilling (in the nervous fibres of vertebrates) really exist in living nervous fibres.

BRIG,[1]) YUNG, KRIEGER, and lately such an eminent authority as LEYDIG. In his work »Zelle und Gewebe« (Bonn 1885) this veteran histologist appears to have changed his view of the structure of the nervous elements, at all events to some extent.[2]) HANS SCHULTZE, HERMANN and those other defenders of the fibrillous structure are, he says, altogether wrong. »Bei Aulocostomum gewähren unter Anwendung der gewöhnlichen Linsen die Nervenfasern den Eindruck einer körnig streifigen Materie. Die jetzt möglichen Vergrösserungen lassen aber finden, dass das »Streifige« von Längszügen eines schwammigen Gerüstes herrührt und das »Körnige« auf die Knotenpunkte eines feineren Zwischennetzes zu deuten ist. Die Haupt- und Längszüge des Maschenwerkesrufen die Abgrenzung in »Fibrillen«[3]) hervor, aber zwischendurch zieht ein zartes Schwamgefüge, in dessen Räumen die homogene, eigentliche Nervensubstanz enthalten ist.«

HANS SCHULTZE's »Primitivfibrillen« are, consequently, according to LEYDIG, only parts of, or longitudinal fibres in, a, usually, rather spongy supporting substance which he calls „*spongioplasm*", whilst SCHULTZE's »Interfibrillärsubstanz«, according to LEYDIG, is the real nervous substance, diffused in the cavities of the spongioplasmatic reticulation; he calls it „*hyaloplasm*".

I think this is enough to show that the discussion which has, now, for 40—50 years been going on, regarding the fibrillous or non-fibrillous structure of the nerve-tubes is not yet finished.

---

[1]) Buchholz describes the cellprocesses and nervous fibres as ribands of a homogeneous substance without any sheath; they are to be considered as »nackte Axencylinder«. — *Solbrig*'s description is very like that of Buchholz: »Allen Nervenfasern der von mir untersuchten Gasteropoden fehlt eine »Schwann'sche Scheide«, und sie erscheinen als hüllenlose Axenfasern mit scharf begrenztem Rande.« . . . »Die Nervenfasern der Mollusken bestehen aus einer festweichen, elastischen, homogenen Masse von schwach glänzendem Aussehen, die in all ihren Eigenschaften mit jener eiweisartigen grundsubstanz, aus der sich die Ganglienzellen aufbauen, übereinstimmt.« . . . . »Durch die Annahme der Nichtexistenz einer Schwann'schen Scheide muss natürlich auch die ältere Ansicht, dass der Inhalt der Nevenfasern aus einer flüssigen Masse bestehe, fallen. Denn eine flüssige Substanz ohne umhüllende Membran ist geradezu undenkbar.« The nerve-fibres have also according to Solbrig a flat form (»die Form von Bändern oder wenigtens von plattgedrückten Cylindern«.)

[2]) This work I have already mentioned on an earlier occasions, vide: Bidrag til Myzostomernes Anat. og Hist. p. 32.

[3]) Hermann's and H. Schultze's »Fibrillen«.

b) The structure of the ganglion cells.

Regarding the structure of the ganglion cells there has been the same disagreement.

A great many writers describe a fibrillous structure, whilst other stick to a homogeneous protoplasm.

Amongst the adherents of a fibrillous structure REMAK comes, also here, first. He describes a concentric striation in the protoplasm occasioned by granular fibres circulating round the nucleus. These fibres do not, however, enter into the processes.

WILL and especially WALTER have also described a fibrillous structure. LEYDIG has in Gasteropodes described a concentric fibrillous structure of the protoplasm of the ganglion cells and a striation of the processes (1865, vide: list of literature. Taf. XIX fig. 3, Ganglienkugeln des Unterhirns von *Helix hortensis*).

Further, SCHWALBE can also be named as an adherent of the fibrillous structure.

BUCHHOLZ supposes the protoplasm of the ganglion cells to consist of two substances, of which the one can, by squeezing, easily be exuded in form of hyaline pearls.

This hyaline »Grundsubstanz«, »in welcher gleichmässig suspendirt ein anderer, in Form feiner Pünktchen erscheinender Körper erscheint«, is »in allen ihren physikalischen und chemischen Eigenschaften auf das Vollkommenste übereinstimmend mit derjenigen, welche die Zellenfortsätze bildet, sowie auch . . . mit derjenigen, welche fibrillär angeordnet den Inhalt der peripherischen Nervenstämme bildet.«[1])

---

[1]) It is really astonishing that so conscientious a writer, as Hans Schultze seems to be, can so seriously misunderstand another author, as he has done, when he says that Buchholz describes »an der Ganglienzellen zwei Substanzen, von denen die eine leicht durch Druck in Form hyaline Tropfen auspresbar, die zurückbleibende, körnige Masse aber durchaus gleichwerthig der fibrillär geordneten Inhaltsmasse der Nervenstämme erschien.« As will be seen, that is quite the contrary of what Buchholz, has really, said. That and similar misunderstandings regarding Buchholz seem, also, to have descended to other writers. In the quite recently published paper by *Rawitz* we find a similar confusion. This writer compares »Buchholz's hyaline Grundsubstanze« with the reticular substance described by himself; a mistake which certainly ought to have been very difficult if Rawitz had looked a little more carefully at the description of »der in feinen Pünktchen erscheinenden Substanz... in der hyalinen Grundmasse gleichmässig vertheilt« given by Buchholz in his excellent paper (p. 252). In reference to the report of Buchholz's paper given by *Vignal*, there is, indeed, little else to say, than that it is thoroughly misleading.

BOLL (1869) says: »Die Ganglienzellen der *Mollusken* bestehen ebenso wie bei den Wirbelthieren aus zahlreichen in den verschiedensten Richtungen verlaufenden äusserst feinen Fibrillen und aus körniger interfibrillærer Substanz. Eine besondere Membran fehlt« (l. c. 1869 p. 19). »Die Nervenfasern, die Fortsätze der Ganglienzelle gehen stets aus der Substanz derselben hervor in der Art, dass die Fibrillen an den Abgangsstellen der Fortsätze eine bestimmte parallele Richtung annehmen und sich zu mehr oder minder feinen Strängen zusammenlegend von dem Zellenkörper abtreten.«

DIETL (1877) describes a concentric striation of the cell-protoplasm, in preparations treated with osmic acid. This striation he supposes, however, also to exist in the live-state. The striation circulates round the nucleus and can always be traced directly into the processes of the ganglion cells. In the brain of *Tethys fimbria* he describes ganglion cells containing a peculiar »protoplasmatisches Balkennetz, das die Verbindung des Zellkörpers und des Fortsatzes vermittelt« (l. c. 1878 p. 524).

CADIAT (1878) does not, exactly, tell whether he supposes the ganglion cells of the *Crustaceans* to have a fibrillar structure in the live-state; their protoplasm is, however, in his opinion, quite identical with the contents of the nerve-tubes, and in the latter he has occasionally observed fibrillæ. He further states that, on application of nitric acid he has seen a striation in the nerve-tubes and »la meme striation se voit sur les cellules et leurs prolongements immédiats.«

The author who has most distinctly declared himself for a fibrillar structure, and who, in my opinion, has made the closest and most convincing investigations on this subject is HANS SCHULTZE (1879). He has defined the point in debate, and has concentrated his investigations upon it, but neither has he been quite successful.

According to the result of his investigations, the protoplasm in the ganglion cells, as well as in their processes and the nerve-tubes, consists of »Primitivfibrillen« and »Interfibrillär-Substanz«; everyone of the fibrillæ is surrounded by interfibrillar substance, they must, consequently, in a manner, swim in it. If that, however, is correct, why then has Schultze, as well as everyone else, succeeded so badly in isolating these freely swimming fibrillæ? why can we only obtain a sort of isolation of them in the extremities of teased processes or tubes, and not in the cell-protoplasm itself?

To this question Schultze has given no reply, nor has anybody else, so far as I know.

FREUD's description (1881) of the structure of the cell-processes is very similar to that of H. SCHULTZE. The cell-protoplasm has however, according to his opinion, another structure, it consists of a reticular substance and a homogeneous one. He says l. c. 1881 p. 31: »Die Nervenzellen im Gehirn und in der Bauchganglienkette bestehen aus zwei Substanzen von denen die eine, netzförmig angeordnete, sich in die Fibrillen der Nervenfasern, die andere, homogene in die Zwischensubstanz derselben fortsetzt.«

KOESTLER (1883) describes the ganglion cells of the frontal ganglion etc. of *Periplaneta orientalis*, as having no membranes: »sie sind nie mit einer Hülle versehen und erscheinen als membranlose, weiche Ballen.« Upon treating the ganglion cells with vapour of osmic acid, and examining them under high powers of the microscope, he has been able to observe »eine koncentrische Lagerung des körnigen Protoplasmas«, »und zwar so, dass das Protoplasma in Schichten geordnet erschien, die rosettenförmig den Kern umgeben« (l. c. p. 585). Of the origin and structure of the processes he says l. c. p. 586: »Eigenthümlich erscheint das Verhalten des körnigen Protoplasmas an der Ursprungsstelle der Nervenfasern. Während sich diese granuläre Zellsubstanz sonst in Schichten koncentrisch um den Kern lagert und der Zelle ein rosettenförmiges Aussehen verleiht, gehen die äussersten Ringe am Ursprunge der Nervenfasern, ihre koncentrische Schichtung verlassend, in diese über. Es ist also dieser Fortsatz nur als eine Fortsetzung des Zellinhaltes aufzufassen, der sogar Anfangs noch etwas körnig, später erst in die feinsten Fibrillen zerlegbar erscheint.«

VEJDOVSKÝ (1884) has not succeeded in observing a fibrillar structure in the protoplasm of the ganglion cells of the *Oligochætes*; »nur die aus den Zellen austretenden Stiele zeigen eine deutliche Zusammensetzung aus den feinsten Nervenfibrillen« (l. c. p. 90). On another occasion he says, however, of the ganglion cells that »bereits ältere Forscher eine fibrilläre Anordnung der Plasmaelemente sichergestellt haben« (l. c. p. 91.)

Amongst those who do *not believe in a fibrillar structure* the following authors may, here, be named[1]):

---

[1]) When speaking of the ganglion cells *Faivre* says »l'enveloppe est solide et consistante; le contenu granuleux, semi-fluide.« *Hæckel* says of the ganglion cells of Astacus: »Jede von ihnen stellt eine mehr oder weniger rundliche, ansehnliche Blase dar, deren zarte Zellmembran, oft von einer dichten Bindegewebskapsel eingeschlossen eine trübe körnige Flüssigkeit enthält in der ein sehr grosser mit einem Kernkörperchen versehener Kern schwimmt.« *Waldeyer* denies the existence of the striation of the cell-protoplasm described by *Walter*. *Buchholz* is already mentioned above.

Owsjannikow describes the ganglion cells of the lobster (l. c. 1861 p. 139) as having »un contenu assez liquide« in which he has observed fibrillæ, »qui semblent pénétrer dans le milieu de la cellule.« The ganglion cells of the *Mollusca* he describes as having a semifluid protoplasm which »erst nach dem Tode fest wird« (l. c. 1870 p. 681). Here, he mentions no fibrillæ.

Cheron (1866) describes (in *Cephalopoda*) apolar, uni-, bi- and multipolar ganglion cells having granular or homogeneous contents. In the »ganglions de bras« he describes, for instance: »des cellules unipolaires à contenu granuleux, et des cellules bipolaires ou tripolaires à contenu liquide sans granules.«

Claparède says (l. c. 1869 p. 593): »Die Zellenkörper bestehen aus einem körnigen Protoplasma ohne erkennbare Membran.«

Solbrig's opinion (1872) was that, the concentric fibrillar appearance was a postmortem one and was, partly, artificially produced by folding of the cell-surface etc. According to his view, the protoplasm of the cells, as well as of the processes, is a homogeneous or granular »fest-weiche« substance. A great many other writers are, also, of the same opinion.

Stieda (1874) says in his description of the *Cephalopoda*, that the protoplasm of the ganglion cells have, in fresh state, an extremely minute granular appearance; »an den Zellen der Schnittpräparate lässt sich eine homogene Grundsubstanz erkennen, in welcher äusserst feine Körnchen eingestreut sind.« The processes have the same appearance (l. c. 1874 p. 92).

Even Hermann (1875) is no adherent of a fibrillar structure of the cell-protoplasm; in his opinion it is rather granulous or homogeneous.[1]

Yung (1878) describes the protoplasm of the ganglion cells as being fluid, and homogeneous or granular.

Bellonci (1878) describes the protoplasm of the large ganglion cells of *Squilla*, as having »un aspetto finamente

---

[1] According to Hermann the protoplasm of the cells consists of two substances: »Die eine umfasst weitaus seinen grössten Theil, erscheint gleichmässig homogen, oder in einigen Fällen feingekörnt, nimmt überall die äusseren Schichten des Zellkörpers ein, gibt ihm seine Gestalt, *und bildet allein den Fortsatz.*« .... »Die andere Substanz erscheint als aus gröberen Körnern bestehend und ist in ungleich grossen Massen um den Kern herum gelagert« (l. c. 1875 p. 30). The processes of the cells have, consequently, in Hermanns opinion a homogeneous structure.

granoso«.¹) Of the peripheral cell-process he says: that it »non presenta alcuna particolarità strutturale«; in Tav. IX, fig. 3 (l. c. 1878) he illustrates a ganglion cell with a peripheral process (»prolungamento periferico«) exhibiting a distinct longitudinal striation which he does not, however, mention in the text so far as I have seen. Some large fibres, he says, have a fibrillar appearance.

NEWTON (1879) designates the cell-protoplasm as being granular; he says that »the granular cell contents may be seen in some instance, extending into the fibres« (i. e. the cell-processes).

KRIEGER (1880) describes a granular cell-protoplasm and a homogeneous process.

VIGNAL (1883) seems to have taken up no distinct position regarding the fibrillar or non-fibrillar structure of the ganglion cells. About that of the *Mollusca* he says: »Elles sont formées d'un globe ganglionnaire à la surface et dans l'interieur duquel se trouvent de fines fibrilles qui forment le ou les prolongements de la cellule; entre les fibrilles se trouvent de fines granulations graisseuses, quelquefois diversement colorées (l. c. p. 342). Regarding the ganglion cells of the *Hirudinea* he reports in somewhat similar terms. »Elles sont formées d'un globe ganglionaire à la surface duquel se trouve un noyau; il est recouvert par de fines fibrilles qui en constituent le prolongement« (l. c. p. 372).

Regarding the ganglion cells of the *Crustaceans* and *Oligochætes* he seems, however, to be of quite another opinion. About those of the Crustaceans he says e. g.: »elles sont formées presque toutes par

---

¹) Of the large ganglion cells he says (l. c. 1878 p. 523): »Esse posseggono una sottile parete e sono circondate da un involglio di tessuto connettivo nucleato. Il contenuto ha un aspetto finamente granoso, ma al polo del prolungamento periferico i granuli sono più fitti e formano un cono distinto, la cui base è la corrispondente parete del nucleo e il cui vertice è l'origine del cilindro assile.« In the small ganglion cells he even believes to have seen »il cilindro assile« penetrate into the nucleus and terminate »in uno spazio chiaro e rotondo che si trova nel centro di questo.« To this apparent connection of the peripheral process or, as I call it, nervous process, with the nucleus, Bellonci ascribes great importance, it shows, in his opinion, that the nervous impression »si propaga al centro della cellula e propriamente al nucleo, il quale nella cellula nervosa, come in tutte le altre, è il vero centro dell'allivita vitale.« We will return to this subject, and to my view of it, at the conclusion of the present paper. Of great interest is, that Bellonci expressly accentuates that each *ganglion cell has only one real nervous process* (»cilindro assile«) »destinato a formare un elemento del nervo, tutti gli altri non sono che prolungamenti destinati ad unire fra di loro le cellule di una stessa massa cellulare.« In the latter statement I do not, however, agree with him (vide sequel).

une substance visqueuse, épaisse, granuleuse et très malléable.« About the Oligochætes he says: »Les cellules nerveuses .... sont formées par une substance demi-liquide visqueuse, excessivement malléable, peu granuleuse. Elles contiennent un noyau réfringent homogène et granulations graisseuses situées à son voisinage.«

If we look through the modern literature having reference to the invertebrate nervous system, and compare the many different views of the structure of the ganglion cells, we meet with a confusion on the subject which is far from encouraging. Some writers distinguish between granulous cells and homogeneous ones, other writers believe in a concentric striation, or even a longitudinal striation (ROHDE).[1]

Some writers distinguish between ganglion cells with a process originating in the nucleolus (»Kernkörperfortsätze«), or nucleus and cells with a process originating in the protoplasm (»Protoplasmafortsätze«). Others, e. g. HALLER, describe cells having both kinds of processes. A great many writers however deny, or doubt, any existence of processes originating in nucleoli or nuclei, etc. etc. All these distinctions and differences of opinion exist, although we certainly must feel inclined, a priori, to suppose that there must be uniformity, to some extent, through the whole animal kingdom in this respect, and that the differences must have arisen in the development of less complicated structures to more complicated ones.

---

[1] In *E. Rohde*'s paper on the *Nematodes* (1885) we meet with a description of the structure of the ganglion cells which is of a somewhat peculiar kind. The writer describes ganglion cells having different modes of striation, a radiate striation, a concentric one, and a longitudinal one (l. c. p. 16—17; fig. 14—34.) As I have not examined the nervous system of the Nematodes I can not, of course, deny the correctness of this statement; if I may judge, however, from the results of my investigations on other animals, I feel inclined to believe that these descriptions are caused, at all events partly, by optical illusion.

It may here, also, be mentioned that *Yung* (1878) describes a longitudinal striation of the protoplasm of the ganglion cells of *Astacus* as being a post-mortem appearance produced by the influence of acids (picric or nitric acid — l. c. 1878, p. 424—425). In the fresh state he describes the cells as having »un contenu liquide absolument identiques à celui des tubes nerveux à l'état frais.«

The American scientist *Packard* gives a very strange description of the ganglion cells of *Asellus* (1884), he says that they »have not, as in the brain of the lobster, a simple nucleus and nucleolus, but they usually have numerous, from 10 to 20, nuclei, the nucleolus of each nucleus readily receiving a stain and forming a distinct dark mass.« How this description is to be explained I certainly can not tell; he does not mention the structure of the cell-protoplasm.

The latest important contribution to the literature on this subject is, so far as I know, to be found in LEYDIG's »Zelle und Gewebe«. According to Leydig's description, the contents of the cells consist, also, of the same two substances *spongioplasm* and *hyaloplasm*, which are mentioned in respect of the nerve-tubes. As in the nerve-tubes the striation — the concentric one in the ganglion cells, and the longitudinal one in their processes — is a rather apparent one, occasioned by »Hauptzüge« in the otherwise reticular spongy *spongioplasm*, through which the *hyaloplasm* is diffused. On a previous occasion he has expressed himself in somewhat similar terms (vide l. c. 1883, p. 56).

In my memoir on the *Myzostoma* (1885 p. 30—31 & p. 74) I describe the protoplasm of the ganglion cells in a somewhat similar way. It consists of the same two substances: spongioplasm and hyaloplasm; the spongioplasm I am, however, »inclined to regard, more, as isolating the hyaloplasm into fibres, than Leydig appears to be.« »The spongioplasm extends, also, into the cell-processes and there, I believe, partly isolates the hyaloplasm into small tubes.« This is, as will be seen, a description very similar to that of Freud, but our opinions regarding the nature of the two substances are quite opposed to each other.

In FRANZ VON WAGNER's work on the nervous system of *Myzostoma* (which appeared at about the same time as my own paper) the author describes the ganglion cells as having a granular protoplasm or also a homogeneous one.

That is, generally speaking, our present state of knowledge regarding the structure of the invertebrate ganglion cells. As will be seen, there are, still, almost as many views as there are writers.[1])

c) The structure of Leydig's dotted substance.

We have mentioned in the foregoing, the history of the nerve-tubes and the ganglion cells of the invertebrates; but there yet remains the most difficult point in debate, viz. the combination of the ganglion cells with each other and with the nerve-tubes, and the real structure of the interposing mass, LEYDIG's »Punktsubstanz«.

---

[1]) In the present review of the literature my attention has been especially directed to the statements regarding the structure of the protoplasm of the ganglion cells, as I take this to be the most important point for my present researches. As to the various statements regarding the existence or non-existence of a cell-membrane etc. and regarding the structure of the nucleus etc. I will, if necessary, refer to them during the description of my own investigations.

Regarding the combination of the ganglion cells with the nerve-tubes, there are two various opinions which have especially been prevalent. According to a great many writers, there is an immediate combination of the cells with the tubes, these being direct continuations of processes from the cells. Another opinion is that, there does not exist any immediate combination between tubes and cells but that the tubes have an indirect origin, i. e., in a fibrillar mass.

The opinion of some authors is, that both modes of origin are present.

A *direct origin* of the *nerve-tubes* in ganglion cells has long ago been maintained by HELMHOLTZ.

Amongst the later adherents of this opinion the following may be named: HANNOVER, WILL, BRUCH, WEDL, FAIVRE, OWSJANNIKOW,[1]) BUCHHOLZ, CHÉRON, BRANDT,[2]) STIEDA, BERGER, YUNG, CLAUS, LANG, SPENGEL, MICHELS, FREUD, KOESTLER, ROHDE, POIRIER.

WALTER, SOLBRIG, BELLONCI, BÖHMIG, HALLER and NANSEN[3]) have

---

[1]) *Owsjannikow* states (1881) that in the *Crustaceans* the nerve-tubes arise directly from ganglion cells. The large longitudinal nerve-tubes are formed by the union of several processes from various cells; and they »forment un système particulier, établissant la relation entre les cellules des noyaux de la chaine ganglionnaire et les cellules du cerveau« (l. c. 1861, p. 136). Of the combination of cells, situated oposite to each other, he says: »Les cellules nervense d'un côté sont partout unies aux cellules de l'autre par des commissures.« The nerve-tubes of the *Molluscs* Owsjannikow also describes (1870), as arising directly from ganglion cells. In most tubes he has, however, observed a division into two branches, the one of which passes into a peripheric nerve, whilst the other one passes »zu der entgegengesetzten Häifte des Nervenknotens«. »Es lässt sich aber auch ferner nachweisen, dass einzelne Gangliengruppen auf ebendieselbe Weise mit einander verbunden sind.« Of interest is, that Owsjannikow (like Buchholz) »an manchen primitiven Nervenfasern auch solche Nebenästchen entdeckt habe, die sich fast plötzlich in eine sehr grosse Anzahl höchst feiner Ästchen dritten und vierten Ranges theilten und endlich so fein vurden, das sie der stärksten Vergrösserung entgingen.« »Diese Ästchen« he supposes to be similar to Deiter's protoplasmic processes, in which supposition I do not, however, agree with him. It may also be mentioned that Owsjannikow, like Walter, describes »multipolare Nervenzellen« situated »an manchen primitiven Nervenästen«. These cells are, however, in my opinion, not ganglion cells but neuroglia-cells.

[2]) *Brandt* (1870) designates the »Punktsubstanz« or »Medullarsubstanz« in the ganglia of *Lepas anatifera* as consististing of feinen Körnchen. Its function or importance he does not, however, mention. The *nerve-fibres* (»welche den Achsencylindern der markhaltigen Nervenfasern der Wirbelthiere und einiger Wirbellosen entsprechen«) *are direct continuations of the processes of the ganglion cells.*

[3]) Lately, a paper by *Rawitz* has appeared, where views are expressed very like those of Haller. This paper will be referred to at the conclusion of the present chapter.

described a direct origin of the tubes in cells, but at the same time have also described an indirect origin as being present, WALTER, SOLBRIG, BÖHMIG and HALLER even suppose this to be the prevalent mode.

WALTER desribes the indirect origin as being produced by interposed multipolar cells, whilst the other writers presume a more or less fibrous or reticular substance as being the interposing medium; this substance is principally formed by processes from the cells.

HALLER and HANS SCHULTZE (as will be mentioned later) describe both modes of indirect origin.

Upon several occasions I, myself, have described an indirect, as well as a direct, origin of the nerve-tubes and supposed both modes of origin to be present to a somewhat similar extent, as, in my opinion, the nerve-tubes having a direct (never isolated as most writers maintain) origin should be motoric ones, whilst those with an indirect origin should be of a sensitive nature.

Amongst those who maintain an *indirect origin* of the *nerve-tubes* from a granular-fibrous mass, and, as a rule, deny the existence of a direct origin, the following writers may be specially named: LEYDIG, WALDEYER, HERMANN, HANS SCHULTZE, KRIEGER, VIGNAL, PRUVOT, VIALLANES, F. V. WAGNER Some of these, e. g. Leydig and Waldeyer, admit a direct origin to occur quite exceptionally.

LEYDIG is the first writer who has given a somewhat detailed description of the central mass of the ganglia. He calls it »Punktsubstanz« and characterises it as a »netzförmig gestrickte Gewirr feinster Fäserchen«. This »Punktsubstanz« receives on one side the branching processes of the ganglion cells (these loose themselves into the fibrous substance) on the other side it gives origin to the peripheric nerve-tubes.

This not very detailed description by Leydig has been supplemented by very few scientists. Most writers seem to be satisfied with it. they use the name without entering more closely upon this difficult subject, and do not try to define the structure of the central mass more exactly. WALDEYER characterises it as a »Gewirr feinster Fäden«, which should originate principally in the division of the cell-processes.

BUCHHOLZ (l. c. 1863) calls this mass »jenes *feinste Fasersystem*. welches überall innerhalb der Nervencentren verbreitet ist.« He has, in my opinion, in many respects, arrived at a very correct view, which is scarcely surpassed by any other writer. The fibrillæ of this »Fasersystem« are extremely slender. »Die Ursprungsweise derselben ist nun, wie angedeutet, eine doppelte; einmal nämlich gehen sie, wie an den multipolaren Zellen zu bemerken ist, aus unmittelbar von dem Körper der Zelle entspringenden, ursprünglich breiten Zellenfortsätzen hervor, welche sich gänzlich in derartige feinste Fibrillen auflösen, das andere Mal dagegen entspringen sie nicht unmittelbar von den Ganglienzellen, sondern werden erst von den breiten Fortsätzen abgegeben, wobei sie alsdann gewöhnlich sogleich als sehr feine Fasern sich darstellen, welche oftmals noch in ganz ausserordentlicher Entfernung von der Zelle selbst entspringen; ohne dass die breiten Axencylinder durch die Abgabe derartiger feinster Reiserchen irgend merklich sich verschmälerten.« — »Diese feinsten Reiser entspringen sehr häufig mit einer sichtlichen plattenförmigen Verbreiterung von der breiten Stammfaser.« These »feinsten Reiser« generally subdivide »so dass aus denselben eine ausserordentlich grosse Anzahl unmessbar feiner Fasern hervorgeht, welche überall mannichfach sich durchkreuzend im Inneren der Nervencentren vorhanden sind.«[1]) These »unmessbar feinen Fasern« are the smallest and finest elements which Buchholz has been able to observe in the fibrous mass, still, however, he is not sure whether they do not subdivide, or if they really are the ultimate branches by which the correspondence between the ganglion cells is produced.

Regarding the appearance of these fibres he says: »dass dieselben je nach der Natur der Flüssigkeit, in welcher dieselben isolirt wurden, mehr oder minder deutlich unregelmässige Varicositäten zu zeigen pflegen.«

The fibres are, as he supposes, separated from each other »durch eine gewisse Menge seröser Flüssigkeit«, which in the preparations have the appearance of »einer fein granulirten Substanz«, the granular appearance is, however, probably artificially produced; perhaps it is, also, to some extent produced by destruction of some of the nervous substance. And he says: »bei dem völligen Mangel geformter Bildungen zwischen der nervösen Elementen lässt sich daher für die Centraltheile ebenso wie für die Nervenstämme als höchst wahrscheinlich annehmen, dass die geringen Zwischenräume, welche zwischen den Gangliencellen und den Fasergebilden übrig bleiben, ebenfalls hier nur von einer des Ganglion durchtränkender Zwischenflüssigkeit erfüllt werden.«

As to the origin of the nerve-tubes, he considers that all of them originate directly in ganglion cells; but their course through the fibrillar mass is, as mentioned, not an issolated one. Each

---

[1]) l. c. p. 289.

cell has, generally, only one process forming nerve-tubes; as a rule this process, however, sooner or later, divides dichotomically; how many divisions there may occur he has not ascertained. »Doch scheint zum mindesten eine dichotomische Theilung stets vorhanden zu sein wenigstens sind mir in mehr bedeutender Länge erhaltene Axenbänder ohne jegliche Theilung niemals begegnet« (l. c. p. 283).

Thus, several tubes must be supposed to originate in each ganglion cell, and really unipolar cells do not exist, according to his view, or if existent they are very rare, and of a small size.

Regarding the form of the ganglion cells (the larger as well as the smaller ones) »so finden wir allerdings den unipolaren Habitus durchaus vorwiegend.« Really multipolar cells are, however, also present; they are generally of a larger size. »Die Fortsätze dieser multipolaren Zellen zeigen nun, ... ein verschiedenes Verhalten an derselben Zelle. Bald nämlich theilen sie sich schon nach kurzem Verlauf gänzlich in ungemein feine Fasern,[1] .... bald verlaufen sie in sehr grosser Länge, ohne sich aufzulösen mit ziemlich gleich bleibender Breite fort, wobei sie nur hin und wieder vereinzelt sehr feine Fibrillen[1]) abgeben« (l. c. p. 276).[2]

According to BUCHHOLZ, an indirect origin of the nerve-tubes, as maintained by LEYDIG and WALTER, never occurs. »Es kommt nirgend vor, dass breite Primitivfasern von feinsten Fibrillen erst zusammengesetzt werden, auch sind sie selbst als homogene Bänder und keinesveges selbst als Bündel feinster Fasern anzusehen« (l.c.p. 305). The function of »jenes feinsten Fasersystem« is, consequently, not to form nerve-fibres with indirect origin, but to produce correspondence between the ganglion cells, which *never* have a *direct connection* with each other as Walter and other writers maintained. »So oft ich aber auch einen derartigen Zusammenhang zwischen Zellen zu sehen glaubte, habe ich mich doch immer wieder in jedem einzelnen Falle davon überzeugt, dass dieser Anschein auf ganz bestimmt nachweisbaren Täuschungen beruhte« (l. c. p. 293).

---

[1]) Cmfr. what is above mentioned regarding the origin of the slender fibrillæ of the fibrous mass.

[2]) According to Buchholz apolar ganglion cells do not exist. About this he says: »Es ist den vorangehenden Arbeiten zufolge kaum nöthig hervorzuheben, dass Ganglienzellen ohne Fortsätze gar nicht vorkommen.« Seeing that Buchholz (and before him Owsjannikow) has already expressed himself so very decidedly on that point, it is really very curious to find this phantom haunting the brains of, even, quite recent writers. Regarding the non-existence of processes issuing from nuclei or nucleoli Buchholz, also, expresses himself very decidedly.

We will close this report of BUCHHOLZ's paper with a reference to his description of some corpuscles occurring in the fibrous mass. About these he says: »Prüfen wir nämlich die feinen Faserzüge, welche neben den breiten Axenbändern überall anzutreffen sind, so bemerken wir an sehr zahlreichen dieser Fasern erhebliche Anschwellungen, welche mit deutlichen ovalen Kernen versehen sind.« »Es sind fast immer langgestreckt *spindelförmige Gebilde*, welche nach beiden Seiten hin in feinste Fasern unmittelbar sich fortsetzen, oder vielleicht richtiger gesagt in den Verlauf derselben eingeschaltet sind.« »Die Kerne enthalten einen oder mehrere sehr feine, punktförmige Nucleoli.« The extremities of the fibres issuing from this »Gebilde« resemble »in ihrem ganzen Ansehen, sowie durch das Vorhandensein vielfacher Varicositäten auf das Vollkommenste den feinsten Verzweigungen der Ganglienzellen.« Besides these »in reichlichster Anzahl vorhandenen regelmässig spindelförmigen Faseranschwellungen giebt es aber noch eine **andere Art** kernhaltiger, mit den feinsten nervösen Fasern in Zusammenhange stehender Bildungen« (l. c. p. 290—291).

These have generally a multipolar shape, and have nuclei of quite the same appearance as those just mentioned. Buchholz does not exactly know which nature he ought to ascribe to these cells, I think, however, there can be no doubt but that these cells are what I have described as neuroglia-cells (vide sequel).[1]

This report of Buchholz's paper is, perhaps, somewhat prolix. I have, however, made it thus circumstantial, partly because I have seen no correct report of the beautiful researches of this eminent investigator, and also, partly, because I consider those researches to be of the highest import, and it really seems as if they are far too little known.

WALDEYER (1863) describes »das molekuläre mittlere Stratum« as »eine Art Flechtwerk, welches eben der Feinheit der Fäden wegen, aus denen es besteht, sehr schwer zu entwirren ist.« This »Flechtwerk«, is formed of the three following elementary parts: »feine Ausläufer der grossen Zellen, kleine Zellen und deren feinste Ausläufer.« »Die kleinen Zellen liegen sehr dicht neben einander« in the central mass. This last description does not suit for my *neuroglia-cells*, which in Mollusca (which Waldeyer has especially adopted

---

[1] It is indeed very strange that *Rawitz* has not observed Buchholz's description of those corpuscles, as he has described and figured quite similar ones (cmfr. the subsequent report of R.'s paper). The multipolar cells of *Haller*, situated in »dem centralen Fasernetz«, are also, undoubtedly, the same cells as those multipolar »Gebilde« of Buchholz.

for his investigations) occur somewhat sparingly, in spite of this, I do not doubt, however, that what Waldeyer has called »kleine Zellen« (»kleine bi- und multipolare Zellen«) is what I describe as neuroglia-cells.

Of the results of his investigations »über den Ursprung der den Axencylindern gleichwerthigen Axenfibrillen der Wirbellosen« Waldeyer, himself, gives us the following summary: »Stellen wir . . . . das Endresultat zusammen, so ergiebt sich: dass dieselben (i. e. »Axenfibrillen der Wirbellosen«) in den von mir untersuchten Fällen direct aus feinen Ausläufern kleiner bi- und multipolarer Zellen[1]) ihren Ursprung nehmen, entweder aus einem solchen Zellenausläufer ohne Weiters, oder so, dass erst Theiläste desselben die Axenfibrillen sind. Niemals aber gehen direct Ausläufer der grossen unipolaren Nervenzellen, welche die Randpartien des Ganglion constituiren, in periphere Nerven über.«

Thus Waldeyer denies the existence of what we would call a direct origin of the nerve-tubes from ganglion cells.

According to CHÉRON (1866) the nerve-tubes of the *Cephalopoda* have, as a rule, a direct origin from ganglion cells, usually in such manner that several processes of small cells unite to form one large nerve-tube. In some cases he supposes that one large ganglion cell alone »fournit directement un tube« (vide l. c. 1866 p. 94). The nerve-tubes of the »système stomatogastrique« have another (indirect?) mode of origin. Of the central mass of the ganglia he seems to have no distinct view; he describes it as »une matière finement granuleuse, absolument amorphe.«

CLAPARÈDE (1869) quite agrees with LEYDIG in his description of the dotted substance. »Die Zentralmasse des Bauchstranges von Lumbricus hat Leydig ganz richtig als eine sehr feine Punktsubstanz mit darin gelegenen dünnen Fäserchen beschrieben. Andere Nervenfasern giebt es nicht, weder im eigentlichen Nervenmark noch in den Nerven selbst. Diese meist geschlängelten Fäserchen verlaufen in den verschiedensten Richtungen die Mehrzahl jedoch der Länge nach; die von den Nerven stammenden strahlen nach allen Seiten in das Bauchmark hinein.«[2]) It is quite evident that Claparède's »Fäserchen«, which he believes to be the real »nerve-fibres«, are the sheaths of the nerve-tubes, which, transsected longitudinally, very often have the »geschlängelte« appearance which he mentions. In

---

[1]) As mentioned above, it is these cells which I call *neuroglia-cells*.
[2]) l. c. 1869 p. 593.

his description of a connective tissue, or »VIRCHOW's Neuroglia vergleichbare Stütsubstanz« in the nervous system, he says, even, that he has not been able to distinguish the fibres of this „*Stützsubstanz*" »von denjenigen der centralen, fibrillären Punktsubstanz .... auch scheinen sie in dieselben unmittelbar überzugehen. Ueber die nervöse Natur der letzteren kann aber kein Zweifel obwalten.«[1]) The mode of origin of these »nerve-fibres« Claparéde scarcely mentions, and it seems as if he agrees with LEYDIG, also, in regard to it.

In his last memoir on the Annelids (1873) he gives a description very similar to what is above quoted.

SOLBRIG (1872) describes the »Punktsubstanz« as a granular fibrous mass which, in a successfully isolated preparation, may be seen traversed by »einem merkwürdig feinen Fasersystem, dessen Fibrillen an dem Rande des Präparates oft auf weite Strecken hin isoliert verfolgt verden können.« He compares this »Fasersystem« with the capillary reticulation of the vascular system, and supposes that a part of its function is to produce the correspondence of the unipolar ganglion cells with each other. Besides a direct origin of the nerve-tubes from ganglion cells, Solbrig also distinctly maintains an indirect origin from this »körnig-faserigen Masse«, the latter mode being the most common one. In this case the nerve-tubes are formed by the union of several »Fibrillen des feinen Fasersystems«.

STIEDA[2]) (1874) supposes the nerve-tubes of the *Cephalopoda* to be direct continuations of the processes of the ganglion cells. Such a relation, he has been able to observe in some few cases. »Ein anderweitiger Faserursprung lässt sich mit Sicherheit nicht demonstriren.« He mentions some »feinsten Nervenfasern, kaum messbare Faden, welche ein im Centrum des Knoten befindliches schwer zur entwirrendes Netz bilden.« His view of the import or function of this »Netz« he does not, however, give.«

The writer who has paid most attention to, and has made the most minute investigations upon, the central fibrous mass of the invertebrate nervous system is, in my opinion, HERMANN (1875). This eminent investigator has, in his really classical memoir, given a very circumstantial description of this substance, describing it as being granular-fibrillous, and in which he very particularly indicates the course and origin of the fibres (l. c. p. 84 etc.).

---

[1]) l. c. p. 595.

[2]) I regret to say that to *Trinchese*'s memoir on the nervous system of the *Cephalopoda* (l. c. 1868) I have had no access, as we do not posess it in the library of Bergen's Museum.

The extremely slender fibrillæ of the mass spring, partly from cell-processes, partly from peripheral nerve-tubes, and partly from the longitudinally and transversally running fibres of the commissures. The fibrillæ unite in »Knotenpunkte« not in such a way, however, as to form a real reticulation.

»Der Zusammentritt zweier Fibrillen verschiedener Herkunft von denen die eine z. B. von den Zellen entspringt, die andere den Commissuren angehört, geschieht in der Weise, dass die eine Fibrille — in Folge der Verlaufsrichtung unter nahezu rechtem Winkel — in die andere unmittelbar übergeht. Nur eine kleine Verdickung die ich oben Knotenpunkt genannt habe, ist an der Verbindungsstelle zu bemerken. Tritt an denselben Knotenpunkt noch eine dritte oder vierte Faser, so verwischt sich allerdings das charakteristische Bild und es ensteht der einer multipolaren Zelle ähnliche Körper, über den ich bereits meine Ansicht mitgetheilt habe.«[1] »Sorgfältigst angefertigte Zupfpräparate bringen jene Knotenpunkte besonders dann gut zur Anschauung, wenn sie nicht zu feinen Durchchnitten entnommen sind. Doch zeigen auch frische Präparate diese Verhältnisse, wobei jedoch die Untersuchung einmal durch die ungemeine Blässe der Fibrillen erschwert ist, ferner dadurch, dass alle nicht vollständig zerzupften und so durch die Mannigfaltigkeit der Verbindungen netzförmig erscheinenden Theile in den Lücken zwischen Fibrillen und Knotenpunkten eine helle, nahezu halbflüssige Zwischensubstanz erkennen lassen, welche durch ihren Reichtum an Körnchen die Deutlichkeit des Bildes beeinträchtigt« (l. c. p. 84—85).

This is quoted, thus circumstantially, because I propose to refer to it in describing my own investigations. From what is quoted, i will be seen that Hermann supposes the central mass to be formed by fibrillæ, and by a granulous, viscous, »Zwischensubstanz«.

---

[1] Although Hermann has never observed nuclei in these multipolar corpuscles, nor in the usual »Knotenpunkte«, he says of the former (l. c. p. 36): »Ich kann diese kleinen Verbindungskörper den übrigen Ganglienzellen zwar nicht gleichstellen, halte für sie aber doch *den Werth multipolarer Zellen* aufrecht, insoferne, als ich sie als Uebergangselemente betrachte, welche die Verbindung zwischen den zur Peripherie ziehenden Fibrillen und den im Obigen beschriebenen grossen Ganglienkörpern, seien es die »unipolaren«, oder die »multipolaren«, vermitteln.« Whether these Hermann's »Verbindungskörper« are identical with the interposed multipolar cells which *Walter* describes. I can not positively deny, but am, however, not disposed to think so. Walter's cells are, I think, identical with *Haller*'s »Schaltzellen«, which I call *neuroglia-cells*, whilst Hermann's »Verbindungskörper« and »Knotenpunkte« are, I think, thickenings of the *neuroglia-filaments* which are also perhaps, to a certain extent, artificially produced by teasing or splitting of the slender tube-sheaths. Their diameter is, he says, .0005—.0006 Mm.

Regarding the origin of the nerve-tubes, Hermann's opinion is, that a direct origin from ganglion cells is very rare, as a rule they have an indirect origin, and are formed of fibrillæ originating in 3 different ways:

1) from ganglion cells. »Diese Fibrillen sind aber *nicht*, wie etwa vermuthet werden könte, *direkte* Ausläufer der Ganglienzellen, sondern entspringen stets von einem *Knotenpunkte*« (l. c. p. 85). On a careful examination of sections this is easily seen. »Die vom Zellvortsatz abzweigende Theil-Fibrille« generally penetrates to the middle of the fibrous mass of the ganglion. Here it unites with »Fasern anderer Herkunft« in a »Knotenpunkt, von dem aus erst die Fibrille in die Bahn der Nervenwurzel, oft unmittelbar neben dem Zellfortsatz, rückläufig übergeht.«

2) from fibrils of the longitudinal commissures, in such manner that a part of these »in Knotenpunkten endet, von denen andere Fibrillen entspringen, um zur Nervenwurzel zu ziehen.«

3) and finally *directly* from fibrils of the longitudinal commissures, without their passing through any »Knotenpunkte«.[1]

In some cases Hermann has been able to detect a direct origin of dorsal tubes of the longitudinal commissures in ganglion cells, the mode in which these tubes terminate he has, however, not been able to decide. Further, he has stated the mutual interpassage of the processes of ganglion cells from one side into the nerves of the other side etc.

Besides what is above quoted Hermann has described a great many interesting particulars regarding the minute inner structure, which, however, it would occupy too much space to mention here. It seems, however, as if later writers have paid much too little attention to these exellent investigations, which, indeed, contain an amount of persevering accuracy, which can not be too highly estimated, and which in my opinion a great many subsequent writers have rather failed in affording.[2]

DIETL (1876, 77 and 78) describes the central mass which he calls »Marksubstanz« as »ein göberes oder feineres, unentwirrbares Netzwerk feinster Fibrillen«. The fibrillæ principally originate in the processes of the ganglion cells. »Die Ausläufer der Ganglienzellen (des Gehirns) tauchen in die Marksubstanz ein bilden hier eine reiche Zerfaserung zum grössten Theile das Substrat derselben und schliess-

---

[1] My view of Hermann's »Knotenpunkte« and fibrillæ will be seen from the description of my investigations on that subject.

[2] The report of Hermann's memoir given by Vignal (l. c. p. 297—298) is, in several respects, quite incorrect and misleading.

lich ordnen sich die Fibrillen neuerdings zu verschieden starken Bündeln aus denen die peripheren Nervenstämme sich entwickeln« (l. c. 1877 p. 24). He does not, however, deny the possibility of a direct origin of nerve-tubes existing. »Ich habe aber unter gewöhnlichen Verhältnissen nur ein solches Verhalten nicht mit untrüglicher Klarheit zur Anschauung bringen können« (l. c. 1878 p. 487).

RABL-RÜCKHARD (1875), in his paper on the brain of the ant, has scarcely paid much attention to the histology of the brain. He calls the dotted substance »jener feinkörnigen, homogenen, keine Zellenstructur zeigenden Substanz, die so vielfach an der Bildung des Centralnervensystems der Arthropoden betheiligt ist« (l. c. p. 489), and his mention of this subject almost confines itself to that.

FLÖGER's paper on the brain of Insects (1878) has scarcely any more interest for our present researches as neither has he paid any particular attention to the nervous elements. In the fibrillar parts of the brain, he distinguishes between masses of »netzförmig gestrickten Substanz« and masses of »längsfaserigen Substanz«, of which he gives no distinct description. The real structure he has not recognised; he says for instance (l. c. p. 561) that the fibres of this »längsfaserigen Substanz«, transversally transsected, have the appearance of points or dots (»bei Horizontalschitten erscheint das Ganze aus zahllosen Punkten zusammengesetzt«). As far as I have seen he does not mention the relation of the nerve-tubes to the ganglion cells; judging from his various descriptions I think, however, that he supposes an indirect origin (i. e. in the central fibrillar substance) of the tubes or fibres, as he would call them, to be the rule.

E. BERGER (1878) maintains a direct origin of the nerve-tubes from ganglion cells; this he has been able to, especially observe in connection with the origin of the *antennal* nerve-tubes in *Musca vomitoria* which he therefore recommends as a good subject for examination to convince oneself of this mode of origin. He does not, however, deny that an indirect origin may possibly occur, and believes it, even, to be probable (l. c. p. 3).

YUNG (1878) believes in a direct origin of the nerve-tubes from ganglion cells. »Les tubes,« he says, »ne sont bien en réalité que de simples prolongements cellulaires.« Still it seems as if he, to a certain extent, e. g. in the origin of the nervi optici, supposes a kind of double origin (direct and indirect?) to exist, this origin is not, however, definitely explained (p. 454). His view of the central fibrous substance seems to be of a somewhat peculiar kind; he describes it as »une substance médulaire, finement ponctuée, divisée en masses plus ou moins cubiques par de fines lamelles conjonctives« (l. c. p. 459). Altogether, his explanations are certainly very indistinct, in another place (l. c. p. 453), regarding the central masses of the brain (he calls it »substance médullaire« a translation of DIETL's »Marksubstanz«) he says: »ils resultent, en effect, d'un complexus de fibres et de substance médullaires que dans ces derniers temps BELLONCI, qui les a retrouvés et déscrits chez la Squilla mantis. a comparé aux grands lobes du cerveau chez les animaux supérieurs.«

HANS SCHULTZE (1879) gives a view of the »körnig fibrilläre Centralsubstanz« somewhat similar to that of DIETL. It has a *reticular* structure with anastomosing fibrillæ; when speaking of it, he uses expressions like »anastomotischen centralen Fasernetz«. He has never succeeded in really observing a direct transition of processes from the ganglion cells to nerve-tubes, but does, however, not absolutely deny its existence. An indirect origin he considers to be the rule, there even exist two kinds of indirect origin: »*centralen Zellfortsätze* lösen sie durch *fortgesetzte Theilung* in die *körnig-fibrilläre netzförmig-verzweigte Centralsubstanz auf*, aus dieser letzteren erst bilden sich direct Nervenfasern. Bei den *Elatobranchiaten* habe ich *wiederholt* einen *anderen Modus* des sogenannten *indirecten Uebergangs* beobachtet. *Kleinen multipolaren Ganglienzellen* sind in die Leitungsbahn *zwischen die grösseren Zellen und die nervöse, anastomotische Centralsubstanz eingeschaltet*. Auch sah ich bei *Gasteropoden* häufig *intercellulare Commissuren*."

This „*anderen Modus*" is, consequently, somewhat similar to what is maintained by WALTER and WALDEYER. As will be subsequently described, these »eingeschalteten kleinen Ganglienzellen are, in my opinion, nothing but *neuroglia-cells* belonging to the inner »connective« tissue or inner neurilem of the nervous system.

CLAUS (1879) maintains a direct origin of the nerve-tubes from the ganglion cells as being the only mode of origin existing in *Phronimida*. »Die peripherischen Nerven wurzeln nicht in der sog. Punktsubstanz, sondern beziehen ihre Fasern aus Ganglienzellen theils des entsprechenden Ganglions — und zwar sowohl gekreuzt als ungekreuzt — theils des vorausgehenden Ganglions, theils vom Gehirne aus.« Regarding the structure of »der sogenannten Punktsubstanz« he is in doubt; »die zarten, als protoplasmatische zu bezeichnenden Ganglienfortsätze« have probably the same relation to this substance in the Arthropods, as they have in the Vertebrates.

»Warscheinlich handelt es sich in der Punktmasse zum grösseren Theile um eine bindegewebige der Neuroglia der Vertebraten vergleichbare Substanz zu der die kleinen ovalen Kerne gehören, welche im Innern der Marklager auftreten.«

NEWTON, in his paper on the brain of the cockroach (1879), has not advanced much further than Flöger in the knowledge of the dotted substance. Under a high power of the microscope, he says that, it exibits a fine reticulation the meshes of which are »extremely difficult to define«. In another part of the brain (the peduncles), he describes a »similar network, but not quite so fine, and the meshes are more elongated (fig. 14), especially towards the upper part, and it is this which gives it a fibrous appearance. It is, in fact, a bundle of fibres which freely anastomose with each other.« From this description, and from the illustration, it is quite evident that the appearance which Newton describes as anastomosing fibres is produced by the sheaths of the slender nerve-tubes, they being transversally or semi-longitudinally transsected. Newton says that »the manner in which these remarkable nervous structures are connected with the other parts of the brain and nervous system has yet to be established.« Neither does he mention the origin of the nerve-tubes or their relation to the ganglion cells. On another occasion, he also mentions a network extending between the ganglion cells. He thinks it, however, to be probable »that connective tissue combines with nervous tissue to produce the appearance presented by their sections.«

MICHELS (1880) has penetrated more closely, into the inner minute structure of the brain and ventral nerve-cord of *Oryctes*. He describes, very circumstantially, the course of the bundles of fibres and cell-processes through the fibrous mass. Though he has not quite understood the real nature of this substance he has, however, a view of it which is more correct than that of many other writers. He calls it »Fasersubstanz« instead of »Punktsubstanz«, »weil ich nach Anfertigung von Längs- und Querschnitten eine molekulären Punktmasse, wie Leydig von den Nervencentren der Arthropoden beschreibt. nicht habe auffinden können, vielmehr immer nur mannigfach sich durchsetzende Längs- und Querfaserzüge wahrzunehmen im Stande gewesen bin.« The greatest part of these »Faserzüge« especially the »Querfaserzüge« spring, in his opinion, from the processes of the ganglion cells; his opinion evidently is also that most *peripheral nerve-tubes spring directly from such processes*. He describes numerous »Querfaserbündel«, »die, aus den Ganglienzellen der *einen* Seite entstehend, den peripheren Nerven der *anderen* Seite bilden, jedoch zusammen mit einem Bündel, das in den Einschnürungen von den jedwede Bauchmarkshälfte durchziehenden drei Längszügen abtritt.« These »Längszügen« do not, however, originate in »einer peripheren Ganglienmasse« nor in »einer centralen Punktsubstanz«, »sondern sind bloss Fortsetzungen jener nach hinten an Dicke abnehmenden Längsfaserzüge, die sich durch die Schlundringskommissuren bis zum Gehirn verfolgen lassen.« They run through the whole length of the ventral nerve-cord.

An interesting view of the central mass is, in my opinion, found in KRIEGER's paper (1880) on the nervous system of *Astacus*. He prefers LEYDIG's name »Punktsubstanz« to DIETL's »Marksubstanz«, finding the former one characteristic, whilst the latter one is misleading. »Die Punktsubstanz ist (he says l. c. p. 540) ein Netzwerk oder vielleicht richtiger ein Filz von feinsten Fasern.« This is easily seen by help of high powers of magnification, in thin transverse sections taken from ganglia treated with osmic acid, it can also be seen in macerated preparations (obtained by maceration in 0.01 % ammonium-bichromate). In such, carefully treated, preparations »wird man deutlich wahrnehmen dass die Punktsubstanz aus äusserst feinen einander durchflechtenden Fäserchen besteht.« Regarding a direct transition of cell-processes into nerve-tubes, he does not deny its existence in *Astacus*, but has, however, not succeeded in really observing it. He supposes an indirect origin of the tubes to be the most common mode. »Die Ganglienausläufer lösen sich, indem sie sich in immer feinere Aeste theilen, in der Punktsubstanz auf oder sie bilden vielmehr dieselbe, indem die durch ihre Theilungen entstandenen feinsten Fasern sich auf die verschiedenste Weise durchflechten, und anderseits kommen die peripherischen Nervenfasern aus den Punktsubstanzballen hervor, nachdem sie sich durch die Vereinigung verschiedener solcher feinster Fasern constituirt haben.« The division is not produced in such a way that »shon vorher getrennt neben einander herlaufende Elemente (Primitivfibrillen), nur ihren gemeinsamen Verlauf aufgeben, sondern der vorher gemeinsame Inhalt einer Faser theilt sich in mehrere Aeste wie sich das Wasser in den Röhren einer Wasserleitung theilt, die die verschiedenen Häuser einer Stadt zu versorgen hat und in ähnlicher Weise verschmilzt der Inhalt der einzelnen Fäserchen die zu einer peripherischen Nervenfaser zusammentreten, nach meiner Auffassung ebenso zu einer gemeinsamen Masse, wie sich das Wasser verschiedener Bäche zu einem Flusse vereinigt.«

J. BELLONCI (1878, 80, 81 and 83) has supplied several beautiful contributions to our knowledge of the central fibrous mass of different invertebrates (Squilla, Sphæroma, Nephrops), as well as vertebrates. He has, very correctly, described it as consisting of connective tissue and nerve-fibres, he has, however, not succeeded in finding the real relation between these two substances: »Selon moi, la substance grenue-reticulée est formée d'un stroma conjonctif et d'un réseau nerveux. Le stroma conjonctif réticulé est excessivement fin dans les parties centrales du cerveau; au contraire, dans la périphérie des ganglions il est plus grossier et ne diffère pas beaucoup de celui dont se composent les capsules cellulaires« (l. c. 1881 p. 178). Through this reticulation of connective-tissue, extremely slender *nervous fibrillæ* run in all directions, anastomosing with each other and forming another reticulation of nervous nature. These fibrillæ are partly derived from processes of ganglion cells, and, partly, they are constituents of peripheric nerve-tubes. He says of them: »ils proviennent *des nerfs périphériques* et *des cellules nerveuses* et se ramifient en une foule de branches grandes et petites qui, les reliant entre eux, forment un véritables réseau.« Besides an indirect origin of the nerve-tubes in this »réseau«, there exists, however, also a direct origin from ganglion cells: »cependant il est certain que beaucoup de fibres périphériques proviennent directement du corps des cellules centrales et, dans les plus grandes cellules j'ai remarqué deux prolongements qui partent du même pôle; l'un forme directement une fibre nerveuse périphérique, l'autre se résout dans le réseau de la substance grenue.« As will be seen, these observations are, in several respects, very similar to mine on the nervous system of Myzostoma and of the Assidians etc. and, also, to many of those, on various nervous systems, which will be described in this paper. Bellonci has found the same »substance grenue-réticulée« in the nervous system of the various animals examined by him.

ARNOLD LANG has, in his various papers (1879, 1881, 1884), afforded some valuable additions to our knowledge of the histology of the nervous system of the *Polyclades, Trematodes* etc. In my opinion, he has formed a very correct idea of the structure of the fibrillar substance in the nerves and ganglia. Of the nerves he says, for instance (l. c. 1884 p. 190): »Das spongiöse Aussehen auf dem Querschnitt kommt dadurch zu Stande, dass der Nerv aus lauter kleinen Bälkchen zu bestehen scheint, welche alle miteinander verbunden sind und welche zahlreiche rundliche, verschieden grosse Lücken umschliessen. .... Auf guten Präparaten aber sind sie (i. e. die Lücken) angefüllt von einer feinkörnigen blassen Substanz,

die an einzelnen Stellen Zellen und Kernen Platz macht . . . .
die blasse, zarte, feinkörnige Substanz ist nichts anderes, als ein
Querschitt einer Nervenfaser. Das spongiöse Balkennetz . . . . . er-
weist sich also als ein Stützgewebe der Nervenfasern. Auf Längs-
schnitten der Nerven ist natürlich von einem spongiösen Bau des
Stützgewebes nichts zu sehen, da die Balken derselben in der Richtung der Nervenfaser ausgezogen sind. In jedem Nerven bildet das
Stützgewebe deshalb mehr oder wenige zahlreiche, miteinander ver-
schmolzenen Röhren, von denen jede eine Nervenfaser umschliesst.«[1])
Of the central mass (dotted substance) of the brain he says, afterwards (p. 190): »Ich glaubte früher, dass sich das Stützgewebe der
Nerven bei den Polycladen nicht ins Innere der Gehirnkapsel fort-
setze, habe mich aber, nachdem v. KENNEL[2]) die entgegengesetzte,
Behauptung ausgesprochen hat, davon überzeugt dass dieser Forscher
im Recht ist.« Lang's opinion must, consequently, be that the spongy
looking »reticulation« in the dotted substance of the brain, as also
in the nerves, is produced by a »Stützgewebe« or, as I call it, *neuroglia*, which in reality forms tubes; an opinion in which I do quite
agree with him, as will be seen from my subsequent description
and also from my previous papers.

In his description of the brain of the *Trematodes*, he even expresses himself quite unmistakably in favour of this view. He
says: »Vergleicht man Schnitte durch das Gehirn mit Schnitten
durch einen der starken Längsnerven, so ist man überrascht von
den Ähnlichkeit der Bilder. Auch im Gehirn treffen wir, wie
TASCHENBERG[3]) richtig bemerkt, auf Schnitten, die in der Längs-

---

[1]) These nerve-tubes, he describes, at first, as anastomosing with each other.
»Die Nerven bestehen aus äusserst zarten, mit einander anastomosirenden . . . .
Fasern« (l. c. 1879 p. 485). Later, in his Monograph (1884), he uses just the same
words, dropping only the expression »mit einander anastomosirenden«. From this
It seems as if Lang has, probably, partly changed his view regarding the anastomosing of the fibres, and if so, he is, I suppose, right if I may judge from my investigations on the nervous system of other animals.

In his description of the nerves of the *Trematodes* he says (l. c. 1881 p. 37):
»In Folge fortgesetzter Theilung solcher Lumina durch neue Scheidewände kommen
die kleineren Höhlungen der spongiösen Stränge zu Stande. Es darf uns deshalb
nicht verwundern, dass wir in den feinsten peripherischen Nervenästchen nicht mehr
das Bild des spongiösen Stranges sondern bloss das einer unregelmässig punktirten
Fläche erhalten; denn hier sind durch wiederholte Theilung der sehr fein gewor-
denen Faserscheiden die Lumina auf eine ausserordentlich geringe Grösse reducirt.«
This is a description which, in my opinion, is also quite suitable for the dotted
substance.

[2]) Vide Kennel l. c. 1879 p. 153.
[3]) Vide Taschenberg l. c. 1879 p. 19.

richtung der Thiere geführt sind, dasselbe spongiöse Gewebe, wie in den Nerven auf Querschnitten. Bei beiden sehen wir auf Flächenschnitten dasselbe System mit einander verbundener Röhren und in diesen Röhren liegen bei beiden gleichartige Ganglienzellen.«[1]

In the »Stützgewebe« Lang even describes what he calls: "*Faserkerne*" (vide l. c. 1879 p. 485 and Taf. XV, fig. 5 and Taf. XVI, fig. 7, fk.); they have completely the appearance of neuroglia-cells. Although Lang does not say anything regarding his view of the nature of these nuclei, I do not think there can be any doubt of their real neuroglia-nature (vide also Monograph 1884, Taf. 32, fig. 9, d, e, f, g, and also Taf. 31, fig. 6, g z 3, g z 4). Lang even describes small nuclei adhering to the processes of the ganglion cells, (»Auch der kleinen, den Ausläufern der Ganglienzellen anliegenden Kerne müssen wir, als allgemein vorkommend, Erwähnung thun.« Mon. 1884 p. 183, Taf. 32, fig. 9, a). These nuclei have, also, quite the appearance of those belonging to the neuroglia, as will be seen from Lang's illustrations. It is, consequently, a situating of neuroglia-cells quite similar to what I have previously described in *Myzostoma*, and to what will be described, particularly of *Molluscs*, in this paper.

In the *Trematodes* there must, in his opinion, be a direct origin of the nerve-tubes from ganglion cells, so far as I understand him. In his description of their brain he says, for instance: »Auf lückenlosen Serien von Quer-, Längs- und Flächenschnitten ist es möglich, alle Einzelheiten des Faserverlaufs zu erkennen und die Fortsätze wenigstens der grössern Ganglienzellen bis in die Nerven hinein zu verfolgen.«[2]

Whether the nerve-tubes of the *Polyclades* have a direct origin in ganglion cells of the central nervous system, or an origin in the central fibrillar substance, he does not mention, and neither does he give any distinct description of the dotted substance in these animals besides what is above quoted (vide p. 53). The central part of the brain, he says, consists »aus einer sich sehr schwach färbenden, ausserordentlich feinfaserigen Substanz, in deren Inneren weder Kerne noch Ganglienzellen vorkommen.« From this description it is, however, evident that we have a similar structure of the

---

[1] Regarding these cells situated in the nerve-tubes, vide foot note 2.

[2] In the peripheral nerves, he describes nerve-tubes wich are direct processes of ganglion cells situated in the nerves. The nerves of the *Trematodes*, he says, (l. c. 1881 p. 37) consist »zweitens aus der Nervenfaser, die, in diesen Röhren eingeschlossen, die Fortsätze der ebenfalls in ihnen liegenden Ganglienzellen darstellt.« As I have not examined the Trematodes, I can, of course, form no opinion of the correctness of this statement.

dotted substance in *Polyclades* as in *Trematodes*, and also in *Molluscs*, *Annelids* etc.

SPENGEL (1881) describes in *Oligognathus* large ganglion cells with processes directly forming large nerve-tubes. LEYDIG's and CLAPAREDE's gigantic nerve-tubes, he supposes to be similar cell-processes. These nerve-tubes have sheaths of connective-tissue, being continuations of the connective-tissue enveloping the ganglion cells. Spengel gives no description of the dotted substance; he supposes, however, the observations just quoted to be of importance for our understanding of this substance. (»Aber auch für die Frage nach dem Wesen der sog. Punktsubstanz werden diese Elemente eine Bedeutung gewinnen müssen.«) I suppose his opinion is, that it is also probably formed by »Nervenröhre« with sheaths — at all events to some extent — and here he is, in my opinion, quite right.

FREUD (1882) does not seem to have paid any special attention to the structure of the dotted substance. The relation of the ganglion cells to the nerve-tubes, he supposes to be the same in *invertebrates* as in *vertebrates*, and he believes, to a certain extent at all events, in a direct origin. He expresses himself, however, very indistinctly on this subject.

VIGNAL (1883) believes only in an indirect connection of the ganglion cells with the nerve-tubes by means of a granular fibrous mass. In his description of the Crustaceans, he says, for instance (l. c. p. 325): »Le centre des ganglions est formé par des fibres nerveuses d'un coté, des prolongements cellulaires de l'autre; ces fibres et ces prolongements se mèlent intimement et forment un plexus d'ou partent les nerfs.«

KOESTLER, in his paper on »das Eingeweidenervensystem von Periplaneta« (1883), mentions the structure of the »Stirnganglion« (Ganglion frontale). Of the minute structure of the fibrillar substance, he says nothing, but that it exhibits »einen netz- oder geflechtartig gestrickten Charakter.« His opinion of the importance of this substance he does not give; he seems to suppose the *direct origin of the nerve-tubes in the ganglion cells to be the rule* when he says: »Ganz deutlich ist der Ursprung der Nervenfasern aus den Ganglienkugeln zu beobachten.« The ganglion cells are, in his opinion, all of them, *unipolar*. Their relation to each other he does not mention, he only says that from them »ausgehenden Nervenfasern gehen nach der Punktsubstanz hin und fast regelmässig so, dass sich die von mehreren Ganglienkugeln ausgehenden Fasern vereinigen und dann gemeinschaftlich in die Punktsubstanz eintreten.«

VIALLANES's voluminous papers (1884, 1885, 1887) on the nervous system of *Atheropods* have not much interest for our present researches, as he has paid no special attention to the minute structure of the nervous elements. Regarding the structure of the dotted substance, he only quotes the general opinions of other authors, and seems, especially, to believe in the descriptions and views afforded by Krieger and Vignal. The *nerve-fibres* (he distinguishes between three kinds: *tubes nerveux, fibres fibrilloides* and *fibres filiformes*) have *no direct origin* in ganglion cells, neither have the ganglion cells any direct combination with each other. Of the dotted substance *(substance ponctuée)* he says for instance (l. c. 1884 p. 14): »C'est d'elle que tous les nerfs tirent leur origine, c'est dans elles que les prolongements des cellules ganglionnaires viennent tous se jeter. Ainsi les cellules ne peuvent communiquez entre elles, les nerfs ne peuvent communiquez avec ces mêmes cellules que par l'intermediaire de la substance ponctuée.«

BÖHMIG (1884) has himself given the following summary of the results of his investigation of the dotted substance etc. of the *Gasteropods*. »Die Markmasse wird von einem Punktsubstanzballen gebildet. Unter Punktsubstanz hat man ein inniges, filzähnliches Gewirr von Primitivfibrillen, aus welchem die Zellfortsätze gebildet werden, zu verstehen. Aus der Punktsubstanz gehen die Nerven durch eine parallele Anordnung der erst wirren Primitivfibrillen hervor. Durch eindringende Bindegewebssepten werden die Fibrillen zu secundären Bündeln vereinigt. Eine directe Uebergang von Zellfortsätzen in die Nerven, also ohne vorherige Auflösung in der Punktsubstanz, kommt vor; und zwar fast bei allen Nerven. Jedoch ist diese Erscheinung selten« (l. c. p. 45). As will be seen, this view is somewhat similar to that of SOLBRIG; the *nerve-fibres* may have a *direct origin* as well as an *indirect one*.

PACKARD (1884) makes, in my opinion, some very interesting statements regarding the structure of the brain af *Asellus communis*. As to the structure and importance of the dotted substance, or »myeloid substance« as he calls it, he does not think himself in a position to say anything with certainty. He says of it (l. c. 1884 p. 6): »This latter substance does not exist in the nervous system of the vertebrates, and just what its nature and function clearly are in the invertebrates has yet to be worked out.« »His own opinion from what little he has seen is, that the myeloid substance is the result of the splitting up into a tangled mass of very fine fibrillæ of certain of the fibres thrown off from the mono-polar ganglion cells, i. e., such fibres as do not go to form the main longitudinal commissures.« As to the origin of the nerve-fibres, and their relation to the dotted substance and the ganglion cells, it will, already, from what is just quoted, be seen that Packard does not quite agree with Leydig; l. c. p. 5 he says for instance: »there is no doubt but that all the ganglion cells give rise to fibres, some of which at least pass directly through or above or around the myeloid substance of the cerebral lobes and form the commissures (i. e., the transverse as well as the longitudinal ones). This independence of the myeloid substance appears to be more general in the Assellidæ, at least this we would infer from Leydig's statements previously quoted.« Packard even thinks it to be »little doubt but that in all Arthropoda certain nerve-fibres arising in the pro cerebral lobes (from ganglion cells) pass uninterruptedly to the last ventral ganglion« (i. e., through the whole central nervous system).[1]) A great many peripheric nerve-fibres, he supposes, however, to originate in the way which Leydig has indicated; he says that in the »myeloid substance« a great many processes from ganglion cells become »broken up into a tangled mass of fibrillæ, which unite finally to form the fibres constituting the nerves of the appendages.« Whether he supposes all peripheric nerve-fibres to originate in this way, or not, I have not understood from his description.

FRAIPONT (1884) affords some interesting informations regarding the histology of the nervous system of the *Archeannelids*. Judging from his description, I think there is no doubt, but that the central nervous system of these primary Annulates has a histological structure which is, principally, of quite the same type as that of the central nervous system of higher Annulates, Arthropods and other invertebrates, to be treated of in this paper. He distinguishes between two constituents, an external layer of ganglion cells and a central fibrillar mass; for the latter he uses various designations: »substance fibro-nerveuses«, »substance ponctuée«, »substance nerveuse fibro-ponctuée«, »mass finement ponctuée et fibrillaire« etc. It is evident that he supposes this mass to consist of a web of slender fibrillæ; l. c. 1884 p. 267 he says, for instance: »Figure 5 (pl. XI), on peut voir au milieu des cellules ganglionnaires les sections trausversales ou obliques

---

[1]) As will be seen in my memoir on Myzostoma (1885), p. 35 & 75, I have expressed myself in very similar terms upon this subject, and have said that I believed similar longitudinal nerve-fibres etc. to be generally present in *Annelids* and *Arthropods*.

de fibrilles très minces mêlées a une substanse finement ponctuée.« Whether he considers these fibrillæ to spring from the processes of the ganglion cells, he does not distinctly say, though I suppose this is his opinion. The ganglion cells, he says, generally send their processes into the fibrillar mass, and here they can sometimes be traced only for a short distance, sometimes he has seen them, even, penetrate the whole fibrillar mass of the ventral nerve-cord (»dans toute son epaisseur«). Whether he supposes a direct origin of the nerve-tubes in ganglion cells or an indirect one, I am not quite sure. In the head of Polygordius he describes two masses of ganglion cells in which »émerge un faisceau de fibrilles nerveuses formant un gros nerf qui constitue l'axe de chaque tentacule« (l. c. p. 263). P. 270 he says: »Beaucoup de prolongements de ces cellules pénètrent dans la masse fibrillaire et contribuent à la formation du faisceau longitudinal; d'autres traversent verticalement la region fibrillaire.« Of the tentacular nerves of Protodrilus and Saccocirrus he says, however, that they take their origin »de la masse centrale fibro-ponctuée du cerveau.« The same he, also, says of the commissures.

VEJDOVSKÝ, in his Monograph on the *Oligochætes* (1884), gives us no distinct description of our subject though he mentions it rather circumstantially. The fibrillar »Punktsubstanz« he describes as consisting of slender nervous fibrillæ. Of the fibrillar mass of the ventral nerve-cord he says, for instance, that it »aus feinen, nur in der Längsrichtung verlaufenden und dicht neben und aneinander liegenden Nervenfibrillen besteht die bald einen gleichen kaum mehr als einen Bruchtheil eines Mikromillimeters erreichenden Durchmesser haben, bald, und dies in der grösseren Anzahl in der Mitte mit einer spindelförmigen Anhäufung eines feinkörnigen Plasmas versehen sind (Taf. VIII, fig. 2'). Weder eine besondere Membrann, noch ein Kern kommen an diesen, wohl den letzten nervösen Formelementen zum Vorschein; indem sie, wie wir später unten erkennen werden, dem Zerfall einiger wenigen ursprünglichen Ganglienzellen ihre Entstehung verdanken.«[1]) This description is written from observations on the living animal. Vejdovský does not mention any interfibrillar substance like Hermann, and others, it is, therefore, very difficult to say whether he has seen the sheaths of the nerve-tubes, and described them as fibrillæ like most writers, or if he has really seen the nerve-tubes, but not observed their sheaths. I feel disposed to

---

[1]) l. c p. 91.

think the latter supposition to be the right one. — The relation of this fibrillar substance to the ganglion cells, and to the peripheric nerves, as also the origin of the nerve-tubes, Vejdovský defines very indistinctly, and I have not, indeed, succeeded in getting any clear idea of his real opinion. Once he says that in *Dendrobœna* he has observed the processes of the ganglion cells penetrate into the fibrillar substance where they subdivide »wiederholtenmalen in die feinsten Fortsätze, um ein merkwürdiges Fibrillennetz zu bewerkstelligen« (l. c. p. 90). »Diese Fibrillen sind jedoch ganz anderer Art als diejenigen, welche in der Längsaxe des Bauchstranges verlaufen. Sie stellen nämlich keine selbständigen Elemente vor, sondern entspringen aus den Ganglienzellen in denen bereits ältere Forscher eine fibrilläre Anordnung der Plasmaelemente sichergestellt haben.«

How this ought to be understood, and what the importance of »diese Fibrillen« is, in Vejdovský's opinion, I certainly can not tell.[1]) Afterwards he tells us, that in other species of *Oligochætes* and especially in the larger ones, he has observed the ganglion cells send their processes directly into the fibrillar substance to form »quer und schräg verlaufende Fibrillenbündel«. The relation of these »Fibrillenbündel« to the nerves he does not mention, neither does he say anything of the origin of the nerves, so far as I have seen; he only tells us a little of the views of previous writers, especially Will, Walter and Waldeyer, and it really looks as if he agrees with these old authors.

SCHIMKEWITSCH, in his paper on »l'anatomie de l'épeire« (1884), has suplied no important adition to the knowledge of our present subject. Like Viallanes, he refers to descriptions, by other authors, of the structure of the »substance ponctuée« the significance of which he states to be very difficult to understand. He supposes the peripheric nervefibres to originate in this substance, and, thus, he does not believe in a direct origin of the fibres in ganglion cells.

REMY SAINT-LOUP, in his paper on »l'organisation des Hirudinées« (1885), agrees with Vignal regarding the histology of the nervous system and has therefore not taken up this subject for his own investigations.

---

[1]) P. 92 he says that of these fibrillæ »sich offenbar nur ein Theil an der Bildung der queren und schrägen Fibrillenbündel betheiligen. Die übrigen feinen Fibrillen schlängeln sich zwischen der fibrillären Substanz und dürften wohl zur Entstehung besonderer Hohlräume wesentlich beitragen.« These »Hohlräume« which he describes as »meist kreisförmige, farblose, undeutlich contourirte Feldchen« are evidently nothing but transsected nerve-tubes.

Of the numerous papers on the invertebrate-nervous system, which have appeared in recent times, we will only refer to a few. The most remarkable work in the latest literature on this subject is LEYDIG's »Zelle und Gewebe« (1885) which has been already mentioned. There, he has modified, or to a certain extent changed, his previous view of the dotted substance. The fibrillar reticulation, which he and others have described, becomes nothing but a spongy net-work of supporting substance, *spongioplasm*, in the cavities of which the homogeneous, really nervous substance, *hyaloplasm*, is diffused; the ganglion cells as well as the nerve-tubes consist of the same substance. The nerve-tubes originate in this way: the spongioplasm, which in the central mass is quite diffusively arranged, unites and forms longitudinal fibres and sheaths which envelope the hyaloplasm like tubes; any isolation of nerve-tubes or fibrillæ in the central dotted substance can thus, as a rule, not arise. It will be seen that this is a rather radical revolution in the views of most histologists. To a certain extent, in respect of the supporting substance of the central mass, it certainly reminds somewhat of Claus; in all other respects, however, Leydig holds a view quite different from that of the author named.

PRUVOT's view (1885) of the dotted substance or »matière ponctuée« seems to be of a somewhat peculiar kind; he has not, indeed, succeeded in finding its real nature. In the brain, he describes it in the following way (l. c. p. 232): »C'est une matière homogène, refractaire aux colorations, offrant un fin pointillé qu' un grossissement suffisant permet de resoudre en petites granulations regulièrement espacées et traversée seulement par quelques rares fibres anastomosées qui proviennent de la substance corticale.« What these »granulations« and »rares fibres anastomosées« really are in Pruvot's opinion, I dare not say, I think it is evident that his powers of magnification have not been sufficiently high.

In his »conclusions générale« (l. c. p. 323) he gives the following summary of his researches on this subject: »En effet tous les éléments fibrillaires nerveux (prolongements des cellules, fibres des nerfs et des connectifs) traversent sans modifications la substance corticale, mais au niveau de la substance médullaire se fragmentent, se résolvent en petites granulations d'abord très rapprochées et disposées en séries linéaires, qui s'espacent peu à peu et se perdent au milieu des granulations voisines pour constituer la matière ponctuée. Cellesi est donc un intermediaire entre la cellule et la fibre, entre l'élément central et l'élément conducteur . . . .« As will be seen, this is no very distinct description; his idea of the dotted substance seems to be somewhat similar to that of Vignal, and also to the old one of Leydig. The nerve-tubes have no direct origin in ganglion cells, but always in the dotted substance. »Ainsi les nerfs prennent toujous leur origine réelle dans la matière ponctuée et toutes les fois qu'ils semblent partir du milieu d'un connectif, qui en est dépourvu ainsi que de cellules nerveuses, on peut être assuré qu'il n'y a là qu'un simple accolement de leurs fibres et qu'il en faut chercher l'origine dans un centre supérieur ou inférieur.«

According to PRUVOT the nerves have two roots. He says (l. c. p. 253): »Chaque cordon (i. e. of the ventral nerve-cord) est luimême divisé en deux et les nerfs de la chaîne y prennent leur origine réelle par deux racines, une antérieure et une postérieure. Les nerfs pédieux étant chez les Annélides incontestablement mixtes par leur fonctions, il ne serait pas impossible que l'une en représentât la racine sensitive et l'autre la racine motrice.«

This is consequently something similar to what HERMANN and others have already described in Annelids. NEWPORT has already described a similar arrangement by Insects 1834 (vide l. c. 1834).

POIRIER (1885) describes and illustrates the nerve-fibres of the *Trematodes* as being tubes with a granulous non-fibrous contents (containing nuclei and bipolar cells), but he does not give any distinct description of the dotted substance of the ganglia. He calls it »une substance spéciale finement granuleuse«. In his opinion, it, however, includes nerve-tubes of which he says: »les tubes nerveus du cerveau et de sa commissure sont remarquables par le peu d'épaisseur de leurs parois et le faible développement de la substance amorphe, qui réunit ces tubes et forme leurs parois.« He seems, thus, to have partly recognised the real nature of the central mass of the ganglia. It seems as if he supposes the nerve-tubes, as a rule, *to originate directly in the ganglion cells.* Of the cells of the brain he says: »Les prolongements de ces cellules nerveuses se continuent directement, soit dans les divers nerfs qui partent du cerveau, soit dans la large commissure transversale qui réunit ses deux lobes, et, de là après avoir parcouru une partie du lobe opposé à celui dans lequel ils ont pris naissance, ils pénètrent dans les troncs nerveux qui en partent« (l. c. p. 605).

BÉLA HALLER, who, half a year later than Leydig, published his paper on the histology of the nervous system of the *Rhipidoglossa*, may, in certain respects, be quoted as a contrast to Leydig. According to his description, the dotted substance — which he calls »das centrale Nervennetz« — consists of a net-work of nervous fibrillæ, which by infinite anastomoses form minute meshes; these nervous fibrillæ spring from cell-processes, as well as from nerve-fibres. If Haller's illustrations of the central »Nervennetz«, are compared with Leydig's of the dotted substance, the resemblance must certainly strike everybody. The only difference is that, Leydig calls his reticulation a spongioplasmic one, in the cavities of which the real nervous substance is diffused, whilst Haller calls his reticulation a nervous one, the fibrillæ of which are surrounded by interfibrillar substance. It may indeed be the same so far, because in both cases there must be a quite diffusive distribution of the nervous substance or nervous reticulation in the central nervous system. The cell-processes which contribute to the formation of this reticulation are, according to Haller, those which do not directly unite with other cells, and which do not directly form peripheric nerve-tubes; the latter he calls »Stammfortsätze«. He even illustrates isolated cells with processes forming a reticulation.

The nerve-tubes have *two modes of origin*, some tubes originate directly in ganglion cells, and are direct continuations of »Stammfortsätze«, other tubes originate in the central »Nervennetz«.

No previous writer has so decidedly and emphatically maintained an infinitely anastomosing, reticular, character in the central dotted substance.

At about the same time as Béla Haller, the writer of this paper published a Memoir on the structure of the *Myzostoma* and, subsequently, a paper on the nervous system of the *Ascidia* and *Myxine*. We have there maintained views very similar to those which will be laid forth here, and will therefore now refer to them somewhat cursorily only. The dotted substance is no anastomotic nervous net-work, but a complicated web or plaiting of nervous fibrillæ or tubes. In transverse sections a reticulation is certainly seen, it is, however, to a great extent produced, by the transsection of the sheaths of tube-shaped fibres traversing, or rather forming, the dotted substance; the reticulation is thus a rather apparent one and is of a „*spongioplasmic*" nature. It is, consequently, the same substance which Leydig has described under this name, but it has not, in the writers opinion, the reticular structure he has ascribed to it.

Each cell has only *one, really "nervous process"*, if the cell has several processes, then the other ones are protoplasmic processes with a nutritive function. The »nervous processes« pass to the »dotted substance«, and, there, they — either quite lose their individuality and sub-divide into fibrillæ, *losing themselves in the fibrillar plaiting* — or they maintain their individuality, and pass through the dotted substance and into a peripheral nerve, *forming a nerve-tube*. They have, however, no isolated course, and give off extremely slender fibrillar branches to the fibrillar plaiting, on their way through the dotted substance. The *nerve-tubes have two modes of origin*, they — either spring *directly from ganglion cells (without isolated course* as above mentioned) — or *indirectly*, from the fibrillar plaited texture.

FRANZ VON WAGNER (1886) who, at the same time as I myself, has described the nervous system of *Myzostoma*[1]) supposes the *nerve-fibres or nerves*[2]) *to have only an indirect origin*; i. e. in the »Punktsubstanz«, which he believes to be »ein dichtes Geflecht feinster Fäserchen, welche aus der pinselförmigen Auflösung der protoplasmatischen Fortsätze der Ganglienzellen hervorgehen. Aus diesem maschigen Filz treten die Nerven heraus.« »Das schwammig-netzige Gefüge«, which Leydig describes, in »Zelle und Gewebe«, v. Wagner has also observed; his opinion of the nature of this substance he does not, however, give.

Towards the close of last year another paper by BÉLA HALLER appeared. In it the author states his results of some investigations on the structure of the nervous system of *Annelids, Arthropods (Tobanus bovinus)* and some *Vertebrates*. He compares these results with his previous description of the nervous system of the *Rhipidoglossa*. In the dotted substance of the latter he found *no connective tissue*, the substance consisted, exclusively, of a central *nervous* reticulation springing from the *processes of the ganglion cells*. In those firstmentioned animals the case is different. In their nervous system the dotted substance is formed by a reticulation of connective-tissue, as well as by a real nervous one.

This is a description very similar indeed to that which BELLONCI already, some years ago, on several occasions, has given of the central nervous system of *Crustaceans* as well as Vertebrates. Haller seems however not to know this Italian author.

In his opinion the nervous system of the *Mollusca* represents, thus, a very primitive state, being difficient in a reticulation of connective tissue, it is in this respect like that of *Coelenterates*. In inverte-

---

[1]) I have previously mentioned the memoir of Wagner in a paper on the nervous system of Myzostoma, which will, I hope, soon appear in »Jenaische Zeitschr. für Nat.« 1887.

[2]) v. Wagner does not believe that the nerves of the Myzostomes are really differentiated into fibres (vide l. c. p. 48.)

brates, as well as in vertebrates, he describes direct anastomoses between the ganglion cells.

Haller also quotes the writer's papers on Myzostoma and Ascidians etc. As they are mostly written in Norwegian, he seems, however, not quite to have understood them. — As will be shown in this paper, Haller's supposition of no connective tissue in the dotted substance of the nervous system of the lower *Molluscs* is incorrect.

d) The combination of the ganglion cells with each other.

Regarding the combination of the ganglion cells, two opinions have especially been prevalent. Either, a direct combination by direct anastomoses is described, and asserted — or the existence of such a combination is denied, the latter view is, strangely enough, maintained by very few writers, and scarcely by any modern writer. The former view is especially defended by WALTER and WALDEYER, and is certainly, without comparison, the most prevalent view amongst histologists, and has existed from a very early period.

Amongst its more modern adherents are HANS SCHULTZE, BÖHMIG, BELLONCI and others; a very zealous adherent is BÉLA HALLER, who scarcely illustrates a ganglion cell which does not anastomose with another. According to this view, or rather theory of a direct combination, there is a prominent disposition in papers on the nervous system to find multipolar cells everywhere; Béla Haller even says, with VIRCHOW, that the more closely the central nervous system is investigated the less numerous will the unipolar ganglion cells be, if they do not indeed quite disappear. The fact is, that multipolar cells and direct anastomoses were necessary to the theory of the combination of the nervous elements and the producing of reflex-movements, and what it is necessary to find to support our theories is very often too easily seen.[1]) As a consequense of this, there are very few writers who have ventured to deny direct combinations or anastomoses between the ganglion cells.

Amongst those who have expressed themselves most emphatically as to their *non-existence*, we may rank BUCHHOLZ and SOLBRIG, who in opposition to Walter and Waldeyer quite decidedly deny any direct combination. The combination between the cells is

---

[1]) It may here be mentioned that the »neuroglia-cells« and fibres, which are interposed between the ganglion cells, have certainly assisted a great many writers to see anastomoses, as they have not understood the real nature of this supporting tissue.

according to both writers produced by the »feinsten Fasersystem« in the dotted substance.

CLAUS (1879) seems not disposed to believe in a direct combination, neither VIALLANES, F. v. WAGNER and others.

The writer of the present paper has, also, on several previous occasions firmly denied a common existence of direct combinations between the cells, not having found any case of indubitable anastomoses between cell-processes.

As the present paper is just about to be completed, a memoir by Dr. BERNHARD RAWITZ on »Das centralen Nervensystem der Acephalen« appears in last volume of »Jenaische Zeitschrift«. As it treats of our subject, I will mention it here.

First, it may be said that the powers of magnification used by Rawitz do not seem to have been high enough, which he also states himself; to this circumstance may perhaps be ascribed some of the results at which he has arrived.

Regarding the structure of the ganglion cells, Rawitz supposes like Buchholz, Hermann, Freud and others, their contents to consist of two substances, »von den der eine eine netzförmig angeordneten der andere eine Zähe unter Umstände ölartige Tropfen bildende Substanz ist, die in den Maschenräumen der ersteren suspendirt ist.« Rawitz's mistake regarding Buchholz's description of the protoplasm of the ganglion cells, we have already before mentioned (note on p. 32). He is not sure whether the reticulated substance is the real nervous one or, »ob man nicht vielmehr die in Tropfen ausfliessende Substanz als die eigentlich nervöse, die netzförmige (Buchholz's hyaline) nur als Stützsubstanz anzusehen hat, will ich definitiv nicht entscheiden, möchte aber die letztere Auffassung der Buchholz'shen vorziehen.« As mentioned already, this comparison with Buchholz is quite misleading, Rawitz has indeed just the same view as Buchholz regarding the hyaline substance, (»in Tropfen ausfliessende Substanze«, Buchholz's »hyaline Grundsubstanz«) which Buchholz decidedly supposes to be the really nervous one (cfr. my report of Buchholz p. 32). As will be seen from my description, I agree with Buchholz and Rawitz in this respect.

Rawitz does not think a fibrillous arrangement in the cell to be very probable.

It is very strange, indeed, that Rawitz seems not to know

Leydig's work: „Zelle und Gewebe", which appeared, at least, one and a half year previous to Rawitz's paper, and is mentioned by Haller in his paper on the Rhipidoglossa; a paper which is quoted by Rawitz. It is so much the more strange, as Leydig in this work describes the protoplasm of the ganglion cells quite similarly to Rawitz.

Regarding the shape of the ganglion cells, Rawitz describes unipolar, bipolar and multipolar cells; *apolar ganglion cells do not exist;* neither has he observed cells with processes originating in the nuclei or nucleoli. The largest cells of all the three kinds occur in the visceral-ganglion, and are motoric, »homolog und analog den Vorderhornzellen im Rückenmark der Vertebraten.«

»Unipolare Zellen im Sinne der alten Histologie« do exist, as well in *invertebrates* as also in the spinal ganglia of *vertebrates*, as previously proved by the same author. That Leydig (1865) denies the existence of such cells is of but little importance; »denn einmal hat Leydig in der citierten Arbeit (vide list of literature 1865) die Ganglien nur in toto, nicht an Schnittserien studiert, und dann leugnen die Physiologen überhaupt das Vorkommen von nervösen Zellen die nicht mit anderen Zellen in *direkter* Verbindung stehen, weil dies mit der Theorie nicht in Einklang zu bringen ist. Darauf aber kommt es allein an: ob es wirklich nervöse Zellen giebt, die mit benachbarten Zellen *nicht in direkter Verbindung* stehen. Ob der Fortsatz sich weiterhin teilt, in ein Geflecht feinster Reiserchen sich auflöst, ob dieses Geflecht mit ähnlichen anderer Zellfortsätze eine netzförmige Verbindung eingeht, ist vollständig irrevelant. Es sei denn, dass man annimmt, jede einzelne Fibrille könnte gleichzeitig zentrifugal und zentripetal leiten. — Nun giebt is aber nicht bloss solch' unipolare Zellen, welche nicht mit anderen in unmittelbarer Verbindung stehen, während der Fortsatz sich in feinste Reiserchen zerspaltet, sondern es giebt auch solche, *deren Fortsatz ungeteilt in den aus dem Ganglion entspringenden peripheren Nervenstamm übergeht.*«[1]) In another place (p. 410) Rawitz also says: »Die Physiologie mag sich sträuben, so viel sie will, sie muss mit der Existenz *wirklich unipolarer Zellen* im Sinne der alten Histologie rechnen.« How much my view differs from that of Rawitz will, I suppose, become evident from my description in this paper (vide sequel), as also from my previous papers on *Myzostoma, Ascidians* etc. What is above quoted is, I think, enough to show into how many dilemmas

---

[1]) l. c. p. 408—409.

the old physiological view of the function and combination of the nervous elements has brought histologists.

Rawitz divides the ganglion cells according to their processes into different kinds, as this, however, in my opinion, is of but little importance I shall pay no further attention to it here. This much I shall only say, that, what he calls »Schaltzellen« — which are multipolar cells having only »protoplasmic processes« (cfr. sequel) and situated only in the central »Nervennetz«[1]) — are in my opinion not at all ganglion cells, but cells belonging to the *neuroglia*.

Rawitz denies the presence of any connective tissue inside the outer neurilem-sheath; this is the more surprising as on the one hand neuroglia-cells, according to my investigations, occur in great number in *Molluscs*, and on the other hand he himself, besides the »Schaltzellen« above quoted, describes (p. 422) and illustrates (fig. 79) some, in his opinion, remarkable corpuscles from the cerebral ganglia of Unio pictorum and Anodonta anatina. He has not been able to understand the significance of these corpuscles; he calls them »geschwänzte Kerne«. I do not think there can be much doubt, but that they are the cells which Buchholz has also mentioned, and which I describe as *neuroglia-cells* in the »dotted substance«. They have, certainly, quite the same appearance and shape.

Rawitz describes numerous direct combinations between the ganglion cells of different kinds. These statements are, however, of but little importance to me, all much the more that Rawitz has not employed for his investigations lenses of very high powers or homogeneous immersion; his observations cannot therefore be considered as quite reliable in all respects, and especially not in these, where the best lenses are decidedly quite necessary, if you are to obtain a position to be able to state anything with certainty.

The processes of the ganglion cells have, according to Rawitz, numerous »variköse Anschwellungen«, from which extremely slender fibrillæ very often issue.

The cell-processes he divides into three kinds according to their course:

1) „*Stammfortsätze*", »die direkt und stets ungetheilt zum peripheren Nervenstamm gehen.«

2) „*Markfortsätze*" which penetrate into the »Marksubstanz«, and are, there, broken up into fine fibrillæ.

---

[1]) »In welches sie eingeschaltet sind, um es in seiner Funktion zu verstärken, und das sie mit ihren Fortsätzen bilden helfen.«

3) „*Protoplasmafortsätze*" which produce the combination of the ganglion cells with each other. They have the same appearance as the cell protoplasm.

The »Stammfortsätze« are rare and only occur in unipolar cells. The »Protoplasmafortsätze« occur in bipolar and multipolar cells as well as also in the smallest unipolar ones, which should, consequently, completely want »Mark«- or »Stammfortsätze«.

In his description of the »Markfortsätze« he expresses himself somewhat indistinctly. »Markfortsätze«, he says p. 416, occur in all cells, besides those which have a »Stammfortsatz«, and this he says after having just described cells wanting »Mark«- as well as »Stammfortsätze«. At one point he says: »Die geminipolen Zellen (consequently bipolar cells) haben nur Markfortsätze«. Immediately afterwards he says: »Keine Zelle hat zwei Markfortsätze, sondern immer nur einen« (l. c. p. 416). How this is to be understood is not, I think, very easy to say.

Immediately afterwards he says, again, that »die *oppositipolen Zellen* (consequently bipolar cells) sind diejenigen Gebilde, welche noch am ehesten als solche betrachtet werden können, welche zwei Markforsätze haben. He is, however, disposed to »auch bei Evertebraten wie bei Vertebraten (cmfr. his previous paper 1883 Arch. f. mikr. Anat. XXI) die oppositipolen Zellen als Nervenzellen sensu strictiori nicht anzuerkennen, sondern sie nur als kern- und protoplasmahaltigen Interpolationen der Nervenfortsätze resp. -fasern zu erklären.« A view in which perhaps I, to a certain extent, can agree with him; though I have only succeeded in finding extremely few similar cells or formations.

Rawitz describes, in the *Acephales*, an arrangement of the cells somewhat similar to what I, on previous occasions, have mentioned in *Ascidians* etc. In the outer layer, unipolar cells exclusively occur, in the middle layer, all kinds of cells, unipolar, bipolar and multipolar ones, and in the inner layer, close to the »dotted substance« multipolar cells principally occur. This is, as I have previously declared, in the closest relation to what I call the nutritive function of the protoplasmic processes, and it has not, in my opinion, the significance which Rawitz ascribes to it.

»Der Markfortsatz oder Hauptfortsatz der multipolaren Ganglienzellen«, of the *Acephales* are, according to Rawitz, »das Homologon des Deiters'shen Fortsatze der polyklonen Ganglienzellen im Vorderhorn des Rückenmarkes der Vertebraten.« And he believes it to be »morphologisch durchaus nebensächlich, ob, wie bei dem Axencylinderfortsatz der Wirbelthiere, ein direkter Uebergang, oder, wie bei dem Markfortsatz der Acephalen, ein indirekter Uebergang zur Peripherie durch Vermittlung eines interponierten Netzes stattfindet.« He believes, thus, the unipolar, »geminipolen« and »pseudobipolaren« cells to be the only ones from which »eine nervöse Erregung ausgehen, resp. in denen sie allein perzipiert werden kann«, whilst the multipolar cells are »Sammelorte für diese Reize« and the »oppositipolen« cells must be considered as »Faseranschwellungen« (vide l. c. p. 422).

Regarding the structure of the »dotted substance«, he arrives at the result, after having mentioned some previous writers: »dass *nur* HALLER die Struktur der Marksubstanz erkannt hat, DIETL dieselbe zu ahnen schien, während alle übrigen Autoren sich mit Bemerkungen wie »Filz feinster Fasern«, »Gewirr feiner Fäserchen« etc. abfinden.« If RAWITZ had known works such as, for instance, that of HERMANN, which he does not seem to know, he would certainly have aknowledged that, also, other writers have tried, and not quite without success, to penetrate into this difficult subject. Rawitz's observations are even in some respects very like those of Hermann.

RAWITZ adopts Dietl's designation »Marksubstanz«, as it is, in his opinion, more characteristic than »Punktsubstanz«, which he even declares to be incorrect; the reason why he does not, however, say. Upon the whole, I think it a very doubtful thing to give new designations where an old and well known one is present and generally used, which is the case here. I think it the more so when the new designation can scarcely be said to be a more characteristic one as to the structure; that it is so as to the contents does not even seem to me to be quite evident. »Marksubstanz« is not used for the same reason that Dietl uses it, viz. because it is situated in the »Mark« or centre of the ganglia, but because it, according to Rawitz, contains »eine nervenmarkähnliche Substanz, welche unter gevisse Bedingungen die characteristischen Erscheinungen des Myelin darbietet.« That it really is »Mark« or *myeline* Rawitz has observed, I do not think is at all proved in his description; why may it not be a substance similar to the hyalin substance of the ganglion cells, and, indeed, I suppose it to be the same, though one, certainly, a priori, may feel disposed to suppose *myeline*, or a similar substance, also to be present, even, if not in the way supposed by Rawitz, i. e. diffusively extended in the meshes of the net-work described by him and Haller.

Of the structure of this central »Nervennetz« he gives a description quite similar to that of Haller. It is a anastomotic network of slender fibrillæ, which is diffusively extended through the whole dotted substance, and which is formed by the »Markfortsätze« of the ganglion cells, as also by the »Schaltzellen«, mentioned above. The only difference between Rawitz's and Haller's descriptions is: that Rawitz ascribes to his »Nervennetz« numerous »Varikositäten«, whilst Haller does not mention anything similar; they occur at the points where the fibres forming the meshes unite; they appear »im mikroskopischen Bilde als dunkle Punkte«; and they are real varioceles and not merely optic products (l. c. p. 429). Besides this »Nervennetz« and the »nervenmarkähnliche Substanz« there is, in the »Marksubstanz«, a third constituent, i. e. slender fibrillæ which, in macerated preparations, may be isolated for long distances, and which are also varicose. »Sie sind ausserordentlich schmal aber nicht smaler als die peripheren Fibrillen. . . . . Sie stellen das Produkt des centralen Netzes dar, aus dem sie sich . . . durch Verschmelzung von 2 oder höchstens 3 Netzfibrillen entwickeln.« . . . »Die Fasern selber treten durch die Maschen des Netzes hindurch, um vom ihrem Bildungsort zur Peripherie zu gelangen und werden wohl durch die markähnliche Substanz von den Fibrillen des Netzes isolirt.« Strange to say, RAWITZ supposes, here, the importance of this substance to be only isolation, whilst the substance he found in the ganglion cells, and which he suposes to be similar to this one, ought, in his opinion, to be considered as the real nervous substance strictly speaking, and the reticulation in the cells a supporting substance; that is consequently quite a contrary view.

According to RAWITZ, connective-tissue occurs neither in the central nervous system of the *Acephales* nor in their nerves inside the outer sheaths, and in spite of this statement he describes cells, occuring in the nerves as well as in the central nervous system, so strikingly similar to those of the *neuroglia*, or *inner neurilem*, that I do not think there can be much doubt about their identity (e. g. comfr. his »Schaltzellen«, »geschwänzte Kerne«, round nuclei and oblong nuclei in the nerves etc.).

Regarding the structure of the nerves of the *Acephales* he says that »die Nervenfasern einfach ein primäres Bündel von Axenfibrillen sind.« This is, in my opinion, a mistake, in which he, however, agrees with HALLER and other writers.

Rawitz wonders that HALLER did not know the paper by BELLONCI on the »Tectum opticum der Knochenfische« (Zeitschr. f.

wiss. Zool. Bd. 35) but at the same time he, himself, does not seem to know the papers by Bellonci on the nervous system of invertebrates, which would seem to be of even higher importance to his investigations.

I think it is, indeed, also very strange, that neither Rawitz nor Haller (nor most modern writers) are acquainted with the exellent papers on the central nervous system of vertebrates by GOLGI; they quote a great many other and less important writers, but they do not seem to know this eminent histologist who, in my opinion, has really introduced a new epoch in our researches into the structure of the nervous system.[1])

If we look back at this review of the literature and compare the statements of the various authors with the results of my investigations, it will be seen that in most respects, and these also the principal ones, I can scarcely agree with any of them; the author to whose views I feel most related is LEYDIG. As to the nerve-tubes and ganglion cells, we have seen that almost all writers, except Leydig, who describe a fibrillar structure suppose the nerve-tubes to consist of *nervous* fibrillæ and *interfibrillar* substance, and in the ganglion cells most of them describe a similar structure, whilst some writers describe a reticulation. Only LEYDIG has decisively expressed himself in favour of the »interfibrillar substance«, hyaloplasm, being the real nervous substance, whilst the fibrillæ should belong to a supporting substance, spongioplasm. As to the dotted substance, we have seen that most writers who have seen a reticulation or fibrillæ etc. in it, have agreed in calling the reticulation a nervous one, and the fibrillæ nerve-fibrillæ. Some writers have certainly described a nervous reticulation as well as a reticulation of connective-tissue, but LEYDIG is the only writer who has decisively said that the whole reticulation was of supporting nature, and that the real nervous substance was the homogeneous one extended in the cavities of the reticulation. Though I do not agree with Leydig, he is, however, the author to whose views my observations are most related, the points in which we essentially differ will be seen in the description of my investigations.

---

[1]) Haller certainly quotes from *Unger* a paper by *Golgi*, this is, however, an old paper (1872) and is relatively of but little importance when compared with Golgi's later publications.

2. The material examined.

For my investigations I have used representatives of different groups of *invertebrates* as also *vertebrates*, I have chosen classes as little related to each other as possible, partly, in order that if I found uniformity in these classes as to the histological structure, the observations might be assumed to be of general importance to the whole animal kingdom, excluding those lower classes which have not, yet, got a relatively more developed nervous system. The Echinoderms and Coelenterates I will pay no attention to in this paper, as I intend to keep them for a separate paper.

As a representative of the lower *Molluscs* I have chosen *Patella vulgata*, it being a large species which I could obtain in abundance in the neighbourhood of Bergen.

Of *Chætopodes* I have examined different species of *Nereis*, as well as, also, species of other families, e. g. Leanira, Nephtys etc.

Of *Oligochætes* I have investigated *Lumbricus agricola*.

Of *Crustaceans* I have particularly investigated the *Homarus vulgaris*, as also, occasionally, *Nephrops norvegicus* and some *Amphipods*.

Of *Ascidians* I have had for my investigations specimens of *Phalusia mentula, Ph. obliqua, Ph. venosa, Ph. prunum, Ascidia scabra, Corella parallelogramma, Ciona intestinalis* and species of *Cynthia*.

Of *vertebrates*, it is especially those from the lowest stage of evolution, viz. *Amphioxus lanceolatus* and *Myxine glutinosa* to which I have paid attention. I have, however, also, examined other vertebrates e. g. *Gadus morrhua, Tinca vulgaris, Rana temporaria, Mus musculus, Felix domesticus, Homo sapiens* and even *Balenoptera rostrata*.

As, however, I intend to write a separate paper on the nervous system of the vertebrates, I will not pay much attention to these investigations on higher vertebrates in this paper, only so much may be said, however, that I, everywhere, have found the most beautiful uniformity in relation to most of the observations reported in this paper.

As to the vertebrates, I shall confine myself to mention some observations on *Amphioxus* and *Myxine*. I got *Amphioxus* in abundance at Naples where, by Prof. DOHRN's extraordinary generosity, I was allowed to study at the zoological station, in the spring of last year in spite of Norway having no working table at the station. Quite recently I received a package from that station containing numerous specimens of Amphioxus most exellently preserved, in different ways, by SALVATORE LO'BIANCO. For this, and many other services, it is a pleasant duty to return Prof. DOHRN my most sincere and grateful thanks.

Myxine I have had in abundance from the neighbourhood of Bergen (Alverströmmen).

---

3. Methods of Investigation.

If my researches, in several respects, contribute somewhat to advance our knowledge of the minute structure of the nervous elements, as I hope they will, and although I, in several respects, have arrived at results very different from most other writers, I think that must principally be ascribed to my methods of investigation, and especially to the methods employed for fixing, hardening, and staining.

I have employed fresh isolated preparations, as well as macerations and sections.

The fresh preparations were, usually, either examined in larger pieces (commissures or nerves) in the blood of the animal as recommended by FREUD (1882) or they were made in this way; the structures (nerves or ganglia), were taken as rapidly as possible from the living animal, and then by help of fine needles[1]) were teased in the blood of the animal, and afterwards examined as quickly as possible. Very often, also, I took as thin sections as possible from the fresh animal, examined them in blood, or teased and isolated the elements from them, by help of needles, in the same fluid. The last method is especially to be recommended for such Molluscs, e. g. *Patella*, as have the pedal nerve-cords and nerves imbedded in the strong muscular mass. A perfect isolation is, however, as a rule, not possible in this way, and therefore *maceration* becomes necessary.

The fluid recommended by B. HALLER is not bad for the purpose. It consists of a mixture, *acetic acid* 5 parts, *glycerin* 5 parts, *Aqua dest.*

---

[1]) For this purpose I would especially recommend glass-needles, which are, I believe, originally recommended by *Stricker*. They are cleaner, smoother and, in several respects, better than any metal-needles. They are, certainly, easily broken, but they are, also, on the other hand, very easily made from a glass-rod, with the aid of a gas-jet. If the hairlike extremity of the needle is passed through the flame before use, a very good point is obtained, blunt enough not to break or cut the elements when cautiously isolated, but still very fine and pointed, allowing the most delicate manipulation.

20 parts. I treated the structures, cut into as small pieces as possible with this fluid for one or several hours (sometimes even for a whole day), then teased them in glycerin (50 %) and examined; or washed out, and stained with *ammonia-carmine* or *picro-carmine;* or diluted *hæmatoxylin* (DELAFIELD's solution) which I, for many purposes, can specially recommend, as it gives a very distinct staining, and the isolation is not very difficult afterwards.

This HALLER's maceration-method is very convenient in many cases, and is an extremely quick one; the isolation is often possible in less than one hour, and in such quickly isolated (or stained and isolated) preparations no change of import in the form or structure of the elements was perceptible.

Another, and for some purposes still better, method is with a weak solution of *Alcohol.* I tried different dilutions from RANVIER's »Alcohol au tiers« [1]) (30 %) down to alcohol of 25 % or 20 % and even 17 % (as recommended by SOLBRIG). I have found the weaker solutions especially very good. The structures — freshly cut sections or pieces (always as small as possible however) — were macerated for one or several days, sometimes even weeks, then stained in *ammonia-carmine* diluted with an equal quantity of macerating fluid, for 24 hours, and teased in *glycerin* of 50 %. *Picro-carmin* and DELAFIELD's *hæmatoxylin* were also employed as staining agents, the latter agent gives, also, here, when employed for a short time, good results.

I usually stain before teasing or isolating, because I think it much more practical, and when one is careful not to employ too strong solutions, and to dissolve or dilute the staining colours in the macerating fluid when possible, it does not at all disturb the facility of isolation in any notable degree.

After having teased the sections or small macerated pieces in glycerin, I always protect them with coverglasses on wax-feet, in order that I may improve the isolation, if not sufficient, by careful knocking on the coverglass with a pencil, or a needle.

Besides the above mentioned methods I have, of course, also

---

[1]) How Rawitz arrives at the conclusion he mentions, that Ranvier's »Alcohol au tiers«, which consequently is about 30 %, preserves less perfectly than the dilution recommended by himself, which contains about 25 % alcohol or less, and how he, in the latter fluid, can keep the preparations for 6 weeks withouth any deformation, whilst in the former fluid fungi grew after a few days, I really do not understand. In my opinion it ought to be just the contrary, if any difference is to be perceived in that respect.

employed the classical, diluted solution of *potassium-bichromate* (0.03—0.1 %). Maceration in this fluid for several days, and staining in *ammonia-carmine* diluted with or dissolved in the macerating fluid[1]) gives very good preparations. A shorter staining with diluted *hæmatoxylin* may also be recommended. Upon the whole, maceration in potassium-bichromate is a method which cannot possibly be omitted, if the most delicate structure is intended to be discovered and explored with good results; it is certainly one of the oldest methods but also one of the best.

By the use of *ammonium-bichromate*, in very diluted solution (0.03—0.1 %), results are, I think, obtained very similar to those obtained by potassium-bichromate. The difference is, at all events, not so great as to make it necessary to dwell upon it here.

A maceration-fluid which is, I think, for many purposes very good, is that originally suggested by LANDOIS, and subsequently recommended and described by GIERKE.[2]) It consist of a mixture of: *chromate of ammonium* (concentrated solution 1 part), *phosphate of potassium* (conc. sol. 1 part), *sulphate of sodium* (conc. sol. 1 part), *Aqua dest.* (20 parts).

The sections or small pieces are macerated for one or several days, then stained with *ammonia-carmine, picro-carmine* or diluted *hæmatoxylin*. The dilution of ammonia-carmine with this macerating fluid, as recommended by GIERKE, is, according to my experience, not possible, as the carmine is precipitated. The isolation by this method is a very perfect one. The more delicate structures are, however, slightly visible, as they become too transparent.

The most important thing in researches upon the histology of the nervous elements is, beyond comparison, to get good sections from well fixed and stained preparations. This is, no doubt, the surest and most decisive method. A very careful preparation is, here, however, of the greatest importance.

A splendid fixing, and at the same time hardening, agent is the *chromo-aceto-osmic acid* (FLEMMING's strongest formula), which for our purpose affords really excellent results. I have employed both of FLEMMING's formulæ and think the strongest one the best. It is

Chromic acid . . 1 % 15 parts
Osmic acid . . . 2 % 4 »
Acetic acid . . . 1 » or less.[3])

---

[1]) As recommended by Gierke (1885, Arch. mikr. Anat. Bd. XXV p. 447).
[2]) l. c. 1885 p. 446.
[3]) *Flemming*: Zeitschrift f. wiss. Microscopie Bd. I. 1884 p. 349.

I treat as small pieces as possible with the fluid (in not too small quantities) for 12 to 24 hours, or sometimes even longer (2—4 days), they are then washed and may now, directly enclosed in paraffin (not *imbedded*), easily be cut, under water or alcohol, on the microtome; if necessary the pieces may also be hardened in alcohol. If cut with a sharp knife, one obtains in this way really brilliant sections. I have even got sections only .005 *mm.* thick. The sections are stained in different ways, *carmine, hæmatoxylin* (DELAFIELD's, WEIGERT's or HEIDENHAIN's solutions) or *coal-tar colours* (eosin, nigrosin, methylen-blue, etc.), or combinations of these different staining methods are used.

As a, for many purposes, really excellent combination, it may be recommended first to stain the sections in an aqueous solution of hæmatoxylin ($1/2$ %) for some hours or longer, then wash out and treat the sections for a day (or longer) in a solution of potassium-bichromate (.5—1 %); afterwards, wash out again (not too much) and stain in DELAFIELD's hæmatoxylin[1]); if overstained, decolour in water to which a few drops of acetic acid have been added. The sections are examined in glycerin (50—100 %) or in canada balsam. The preparations obtained in this way are, indeed, in many respects superior to any others.

As however the method is certainly somewhat complicated, I prefer, when possible, to stain in toto. This affords very good results when the pieces are sufficiently small, and are treated carefully. In this case, of course, a longer time is necessary for staining, depending on the size of the pieces. DELAFIELD's *hæmatoxylin* I generally employ very diluted, for staining in toto.

Another method, which affords excellent results for some purposes, is fixing and hardening in alcohol gradatim. Especially for Annelids, it is very good first to narcotise the animals, by carefully pouring a thin layer upon the surface of a small portion of seawater in a glass in which they live. When narcotised they are stretched on a waxplate and hardened in alcohol gradatim. When sufficiently hardened, (not too much) they are stained in a watery solution of hæmatoxylin, and afterwards treated with potassium-bichromate as originally recommended by HEIDENHAIN. For some purposes fixing in a *saturated watery solution* of *picric acid* is still better, then washing out in diluted alcohol and staining as above,

---

[1]) As will be seen this method is very like that already recommended by Flemming, for staining of glands. Zeitschr. f. wiss. Microscopie Bd. II. 1885 p. 517.

— or, also, fixing in LANG's *fluid* (corrosive sublimate 12 % in seawater, or in an aqueous solution of chloride of sodium 6 % acetic acid 6 %, alum 0.5 %) then washing out, hardening a little in alcohol gradatim and staining with HEIDENHAIN's hæmatoxylin as above.

All these methods can be safely recommended for trials; everyone of them will, I think, for certain purposes afford results which certainly are obtained by none of the methods employed before.

My list of methods is however not yet finished. For some purposes even the above mentioned ones were not sufficient, e. g. in *Mollusca*, where it was extremely difficult to arrive at any clear idea of the most minute structure of the nerves and LEYDIG's »Punktsubstanz«. The following method gave, however, excellent results.

The pieces for examination, cut as small as possible, were treated with *osmic acid* (1 %) for 48 hours, then washed in running water, and cut at once by the hand or in the microtome or they may first be hardened in alcohol and then cut. The sections, transverse and longitudinal, were stained in DELAFIELD's *hæmatoxylin* (diluted), and decoloured in water containing a little acetic acid. The sections were examined in glycerine or canada balsam. In this way, very distinct preparations of the fibrillar substance are obtained; the substance obtaining a distinct blackish staining.

Finally, I shall now mention a method whose importance for our future knowledge of the nervous system can scarcely be overestimated, as it affords really quite marvellous preparations and far surpasses every method hitherto known.

This is the black *chromo-silver method* of Prof. GOLGI (at Pavia). By modifications of this method I have obtained exellent preparations, even from the spinal nerve-cord of Fishes, in which nobody before has succeeded. Dr. FUSARI, assistent at the histological laboratory in Pavia, told me that he had worked for more than a year with the nervous system of fishes without getting any staining by this method in the spinal nerve-cord. In the brain, however, the method gave excellent results, which indeed Dr. FUSARI's preparations also richly prove. This gentleman is, so far as I know, the only one who, besides myself, has successfully up to this time employed the method for fishes. I am largely indebted to him for the communication of his experiences obtained in a lengthened use of the method.

Besides on fishes and other vertebrates, I have also tried the method on several invertebrates, not yet, however, with so much success, as I think may, ultimately, probably be obtained. Still, I have also obtained staining here in several groups of invertebrates and, in further experiments, it may probably be possible to find a modification of the method, by which it would succeed on a larger scale. In *Homarus* I have already obtained very good results; as communicated a year ago[1]) I have tried it on Ascidians, with some success. In Mollusca, I have also seen so many signs of a beginning reaction that I feel convinced that it is possible to obtain good staining, if only the most suitable modifications are employed. This perhaps a near future may enable us to succeed in.

As I am still experimenting on this subject, I shall, for the present, only quite shortly communicate the method I at present employ for the spinal nerve-cord of *Myxine glutinosa*.

The *nerve-cord* is cut out of the living animal. It must not be *isolated*, but must, necessarily, be taken with the surrounding sheaths, muscles and corda spinalis if any reaction at all is to be obtained. It is, however, also necessary to take care that the surrounding mass is not too thick, because in that case the hardening agents will penetrate too slowly; the half thickness of the corda spinalis may, therefore, very well be removed. This done, it is desirable to cut the preparation into short pieces (one or a few centimetres long) or, if one wishes to keep it in longer pieces, then, to make, in these, deep incisions with a sharp knife at short intervals. This done, the preparations are laid in a solution of potassium bichromate (2—2.5 $\%$) for about an hour, then the solution is changed and made a little stronger (up to 3 $\%$ or more). Here they remain for about 24 hours; if there is plenty of solution it is not generally necessary to change it again. After 24 hours the preparations are put into a new solution consisting of 4 parts of 3 $\%$ solution of *potassium-bichromate*, and 1 part *osmic acid* (1 $\%$ sol.) in this solution they remain for about three days (72 hours); if there is not sufficient solution, it is necessary to change it after 1 or 2 days. Sometimes I have also employed solutions with more osmic acid (1 part osm. ac. [1 $\%$] and 3 parts potass.-bichr. [3 $\%$] as recommended by GOLGI) sometimes also with less osmic acid (1 part osm. ac. and up to 6 or 7 parts potass.-bichr.) The good results of these different solutions depend much upon the temperature maintained in the rooms where the preparations are kept. After three days, or about that time, (the duration must be tested by results) the preparations are directly treated with *silver-nitrate*. At

---

[1]) *Nansen* l. c. 1886.

first it is well to wash them in a weaker solution (.5 %) of silver-nitrate and then afterwards place them in stronger solutions (up to 1 %). After one day the staining is generally complete. If one wishes, however, to keep the preparations for some time before cutting them, they must be kept in a clean solution of silver-nitrate. It is best, of course, to keep them in opaque bottles; kept in that way they are not destroyed even for months. If we wish to take sections we may cut them, directly from the silver solution, with a sharp razor, under alcohol. The sections need not be very thin; when cut, they may be, preliminarily, examined at once in glycerin, if the staining proves to be a good one — i. e. if ganglion cells with all their processes, and nerve-tubes with their ramifications, appear quite dark or black on a transparent field, — then new sections are made and washed well in alcohol of about 90 or 96 %.

This washing I have usually performed in the following way: the inferior end of a funnel with a rather wide tube is closed with a plug of cotton so as to form a kind of filter. The sections are placed upon the cotton, then the tube of the funnel is filled with alcohol and another plug of cotton is pushed down the tube to a certain distance above the sections; the sections are thus situated in a small tube-chamber filled with alcohol. When now, however, the body of the funnel is filled with alcohol, a stream will slowly filter through this chamber and thus the alcohol in it will be constantly renewed. By putting more or less cotton into the tube one may regulate the velocity of the stream through the chamber. When the sections are sufficiently washed (in 4—8 hours) they are placed in *absolute alcohol*. If there are many sections, the alcohol is changed once or twice. Then (after some hours) the sections are placed for some time (some hours or more, even a day) in pure *turpentine* which has to be changed several times. Then they are placed on the slide in dammar-resin dissolved in turpentine and *protected by no coverglasses* if you wish to keep the preparations for a long time. The dammar is at once dried in a warmbath or in an incubator, where the turpentine is very rapidly evaporated and the dammar becomes quite hard and smooth; a coverglass prevents the turpentine and other volatile oils evaporating. If the dammar is not quite smooth in some places after drying, a little more dammar is added to these places, and the drying repeated. A very good method of mounting is, of course, that recommended by Prof. GOLGI, it is, however, a little more complicated. Prof. GOLGI mounts the sections, in dammar, on coverglasses, the coverglasses are again mounted on wooden slides, in the middle of which square apertures are cut to suit the glasses.

This is an exellent method and admits of the employment of oil immersions, another advantage is, also, that the sections can be examined from both sides, which is often of great importance when the sections are thick. For many purposes the above mode of mounting on common glass-slides will however do. The preparations ought to be kept in darkness when not used. Series of sections stained by this silver method were obtained in the following way.

A piece of the spinal nerve-cord was disected out from a stained preparation, and treated quite in the same way as above indicated for the sections: washed in alcohol, transferred to abs. alcohol, then to turpentine. The only difference is that, I let the stream of alcohol pass quicker over it. From turpentine it is transferred to a solution of paraffin in turpentine which is placed in an incubator heated to 56° C. and to which solution paraffin is again added, then into pure paraffin and then imbedded. The sections are now made at once, then fixed on the slide or coverglass by collodion, and mounted in dammar in the above indicated ways. This method may be recommended also for general purposes, even where no series are required, as it is much quicker and more convenient than the other method where each section must be treated and mounted separately, though it certainly does not afford sections with distinct staining of such permanency.

If convenient, the sections of silver-stained preparations may also be stained with colours dissolved in alcohol, e. g. eosin, safranin, methylen-blue etc. In this way very nice looking preparations may be obtained.

There are probably a great many ways in which this really exellent method of GOLGI may still be improved, and I would recommend it to every histologist of the nervous system, for further experiments.

Description of my investigations.

I. The structure of the nerve-tubes in invertebrates.

Homarus vulgaris.

The nerve-tubes of the lobster *(Homarus vulgaris)* consist, as is well known, of an external *sheath*, with nuclei, and a viscous *contents*. They have a rather homogeneous and refractive appearance with doubly marked outlines when they are examined in the live-state under low powers of the microscope.[1]) On application of higher powers, a longitudinal striation, not very prominent but still quite distinct, is seen to form an axis in the centre of some large nerve-tubes (vide Pl. II, fig. 12, A & B). This striated axis is, in some tubes, very narrow (fig. 12, A) and in the rest of the tube-contents no striation is visible, but only a few lines close to the axis. In other tubes the striated axis is broader, and more numerous lines are seen in the contents round it. In some tubes the striated axis is even very broad (fig. 12, B), and lines are visible almost through the whole contents except in the peripheral layers; they do not usually occur there or are, at all events, very rare. Upon a careful examination of several preparations it is, in fact, possible to find every stage of striation, from tubes with a striation through the whole mesial parts of their contents, to tubes with a very narrow striated axis in their centre, in all these tubes it is, however, possible to observe a concentration of the striation towards the centre of the tubes, in such

---

[1]) For examination of the tubes in the live-state, I think it is best to take long pieces of the nerves or commissures with their sheaths, as intact as possible, and examine them quickly in the blood of the animal; a method which is especially recommended by Freud (l. c. 1882). The oesophageal commissures I found to be very suitable for the purpose as they are not too thick and can be very quickly extricated. I have, of course, also isolated the tubes in the blood of the animal, and think this method to be very good, for some purposes, when it is quickly performed.

manner that the intervals between the lines are narrower in the centre than they are more peripherically, and this is all the more prominent the narrower the axis is.

This striation in some large nerve-tubes is already observed and described by REMAK, and subsequently to him a great many other writers have described it, even HÆCKEL, who describes the tube-contents as being homogeneous, has seen it (as mentioned p. 30 foot note 4) in some tubes. FREUD assures us, however, that a similar striation is visible in all nerve-tubes in a quite fresh state. In spite of this statement of FREUD's and in spite of the application of the best lenses (Zeiss's new apochr. lenses) I must confess that, I have found it extremely difficult to observe a striation in a great many nerve-tubes. In the greatest number of the slender tubes and tubes of middle size, and even in some large tubes, it required, certainly, a predisposition to see a striation, and I really think that an uninterested eye would, in many cases, see no striation, even when the nerves or commissures were instantly taken from a quite fresh and living animal, which FREUD states to be a quite necessary condition if a striation is to be seen in all tubes. I have also, in some rather slender tubes, observed a but little prominent striation (fig. 12, *b*) but there are a great many tubes and, as mentioned, even large ones which have left me very doubtful as to their striation in the fresh state.

I do not think this, however, to be decisive as to the fibrillar or non-fibrillar structure of these tubes in the live-state, because, a priori, I think it to be very probable that if there are two substances in the tubes, their refractive difference is perhaps so small that, when one of the substances is present only in very slender filaments or membranes, the whole contents look as a homogeneous mass, and that the refractive difference must be altered by chemical (fixing) agents if the substances are to become easily visible. That such is the case, and that a distinct striation becomes visible in all nerve-tubes when treated with various fixing agents (chromic acid, osmic acid, picric acid, acetic acid, nitric acid, potassium-bichromate, etc. etc.) is a so well-known fact that it, certainly, need not be further mentioned here. (Pl. II, fig. 15 & 16 epresents some smaller nerve-tubes; macerated for some hours in glycerine, acetic acid and water; a striation was distinctly visible.)

Almost all writers who have seen a striation in the tubes in the live-state have agreed in declaring the darker thin lines of the striation to be nervous fibrillæ (»Primitivfibrillen«) swimming in an interfibrillar substance, this also explains their anxiety to find a striation in all tubes. As will be seen from my present paper, I do not agree with any of those writers, and when the further results of my investigations are known I hope we shall find a reliable explanation of the reason why a striation is not more generally visible than it is.

LEYDIG forms an exception to most other writers in declaring the dark lines of the striation to spring from a structural support, *spongioplasm*, and not from nervous fibrillæ (vide Leydig l. c. 1885, mentioned before, p. 31). He has no doubt arrived at a more correct view than any other writer, but neither is he, in my opinion, quite correct, as will shortly be seen.

If the view that fibrillæ and interfibrillar substance are present in the tubes, was correct, it ought, in my opinion, to be easy to isolate these fibrillæ by splitting the tube-sheaths longitudinally. This is, however, very far from being the case. I have split large and small nerve-tubes in the fresh state longitudinally and no fibrillæ of the kind were visible, I have teased them very carefully and some thicker or thinner filaments became visible in the extremities of the tubes. I never succeeded in obtaining, in spite of my most persevering care, neither fresh preparations nor macerated ones with brushes of regular fibrillæ in the extremities of the tubes similar to those illustrated by H. SCHULTZE in *Molluscs* and *Annelids*. The filaments or fibrillæ visible at the extremities of the nerve-tubes had always a somewhat irregular appearance, with varying thickness and length, and they certainly looked as if they belonged to a supporting substance and were to a certain extent artificially produced.

A viscous homogeneous substance, as stated by SCHULTZE, was certainly also visible between those filaments. This substance never occurs, however, in such a way that I can say it formed an interfibrillar substance, and was diffusively extended in the nerve-tube. At the extremities of the nerve-tubes it always appeared on pressure in the form of regular small pearls issuing from the tube-contents. I never saw it appear in large pearls, not even in the large nerve-tubes where this substance, according to a great many writers, forms a thick layer surrounding a central bundle of fibrillæ. It had, in fact, always the appearance of these small pearls of viscous substance issuing from a great many extremely slender tubes or channels contaning the substance, situated close together and forming the whole contents of the nerve-tubes, and this we shall see is really the case. The filaments visible at the extremities of the nerve-tubes gave always the impression that they could have been produced by the sheaths of those slender tubes being split longitudinally and thus forming fibrillæ. The viscous substance is certainly very often seen adhering to these fibrillæ, but if we look carefully we will see that it always is in the form of small pearls adhering to the sides of them; I have never observed pearls quite surrounding, or embracing, a filament, which certainly ought to occur occasionally if the filaments were round fibrillæ swimming in an interfibrillar substance in the nerve-tubes.

I have not been able to obtain any further information regarding the structure of the tube-contents from fresh preparations.[1]

If we examine *macerated preparations* we will not become much wiser, and neither do they suffice to solve the riddle as to the real nature of the fibrillar structure.

The quickest and easiest method of maceration is that indicated by BÉLA HALLER (cmfr. p. 73—74). In that way an isolation of the tubes is possible even after a lapse of one or two hours; the tubes exhibit a distinct sheath, with sheath-nuclei, as well as a distinct longitudinal striation of their contents (fig. 15); a staining is not necessary. In successful preparations you may find nerve-tubes with extremities more fibrillar and brush-like than any of those obtained in fresh preparations. With a good will and skilful hand you can, by teasing with fine needles, even improve this fibrillar appearance, nay, you can split not only the tube contents but also the sheaths longitudinally into filaments or »fibrillæ«. This shows, however, that these fibrillæ can be artificially produced, because most writers agree in declaring the sheaths to be homogeneous membranes of connective substance *(neuroglia)*. If we now examine the fibrillæ under high magnifying powers we will find them all to have the same, somewhat irregular appearance both, those from the tube-contents as well as also those springing from the tube-sheaths. If we try other maceration fluids (e. g. weak sol. of alcohol, amm. bichr. 0.03 %, potass.-bichr. 0.1—0.03 %, LANGDOIS's fluid etc.) we obtain very similar results. By all these methods we arrive at the

---

[1] In Pl. II fig. 13 an appearance of the nerve-tubes is illustrated which I have often noticed in the fresh state, and which easily could give rise to serious mistakes. *a, a* are the hyaline-looking contents of the nerve-tubes (belonging to a peripheral nerve); *b, b* are the sheaths from which septa (*c, c, c'*) aparently issue and penetrate into the contents of the tubes. These apparent septa are, however, an optic illusion only artificially produced by a slight bending of the nerve in which the nerve-tubes are situated.

conviction that a fibrillar structure is present but how this structure is to be regarded, that is the point respecting which we remain in uncertainty.

There remains then only our last but at same time best resource and that is to try what information we can obtain from transverse and longitudinal sections taken from preparations fixed, stained, and hardened in the best way our present histological technology can afford.

As a very good method, for this purpose, I would especially recommend the following: the nerves or commissures, taken directly from the living animal, are treated with FLEMMING's *fluid* (chromo-aceto-osmic acid, strongest solution cmfr. p. 75—76) for 12 to 24 hours, then washed for some hours in running water, and stained with alum hæmatoxylin (DELAFIELD's formula), or in an aqueous solution of hæmatoxylin (0.5 $^0/_0$), and afterwards treated with a solution of potassium-bichromate (0.5—1 %) as recommend by HEYDENHAIN (vide p. 76). Now the structures are washed, again and hardened, somewhat quickly in alcohol gradatim. Preparations treated in this way can, as a rule, be very easily cut into thin sections, directly enclosed in paraffin only (vide p. 76). Embedding in celloidin or paraffin can of course also be employed; this can not, however, be performed with too great care as we have here to deal with structures of the most delicate nature.

In preparations carefully treated in this way, the nerve-tubes, with their contents, are fixed in their natural position and state, and no visible change of the form has taken place.

If we now examine good and thin longitudinal sections, it will already, under low powers of the microscope, be easy to observe a distinct longitudinal striation in the larger as well as the smaller nerve-tubes. The extremely slender longitudinal lines or »fibrillæ« have got a distinct blackish staining, and are situated with intervals between them, just similar to what we have found in the fresh state. The substance in these intervals is not stained, and has a homogeneous or rather a slightly granular appearance. If now, however, the stained fibrillæ are real fibrillæ swimming in this homogeneous unstained substance, it is evident that they, in thin transverse sections of nerve-tubes, must appear as minute black dots or points, just similar to what, VIGNAL for instance, has also really described and illustrated.

But what do we find? — *Instead of black dots, we find in the sections of the tubes a delicate reticulation, with minute circular meshes, apparently formed by extremely slender filaments* (fig. 2; fig. 5, *t*).

In these filaments we certainly find small dark granules (fig. 6, *a*) which, occur however, only in the point or knots where several filaments unite; they never occur in the centres of the meshes. And the same reticulation is extended through the whole transverse section of every nerve-tube, and fills the whole space inside the tube-sheat.

If we, now, compare the impressions which transverse sections of the nerve-tubes give, with those obtained by longitudinal sections, there can, in my opinion, be no doubt left but, that *the „fibrillæ"* of most writers *belong to a substance, spongioplasm, forming a bundle af slender, cylindrical tubes or channels enclosed in the neurilem-sheath of each nerve-tube. These primitive tubes,* if we may call them so, *are filled with the homogeneous, viscous substance, hyaloplasm, which we already know from fresh nerve-tubes.*[1]) The granules in the transverse sections show themselves to be transsected thicker longitudinal fibres of spongioplasm which, occur especially along the longitudinal edges where several tubes, usually three, meet. If we imagine the spongioplasm as forming cylindrical tubes laid or pressed together, it will be evident, that the spongioplasmic walls of the different tubes will unite and form septa, as the spongioplasm cannot, of course, be considered as a quite solid and unadherent substance; it will also be evident that, in the corners where several tubes meet the spongioplasmic walls will be still thicker.[2])

In this way, I think, we may easily understand why it was so extremely difficult to isolate and get a clear idea of SCHULTZE's »Primitivfibrillen«. As we have seen, they do not exist in the way he has explained; what he called fibrillæ, are the spongioplasmic walls between the real »primitive fibrillæ« or *primitive tubes* as I have called them.

The above description refers to the structure of nerve tubes, which are, I think, in the most primary state.

There are, however, in the longitudinal commissures, as well as also in the peripheral nerves a great many tubes exhibiting a kind of concentration towards an axis in their centre. — As described p. 81, a more or less concentrated longitudinal striation was visible in the

---

[1]) The primitive tubes have a diameter of about .0015—.0017 Mm.; their size vary, however, and is very difficult to measure.

[2]) There occur, however, besides these granules larger dark granules, as well in transverse sections as in longitudinal ones. They are usually very sparingly spread, especially near the centres of the nerve-tubes, as will shortly be mentioned and hey are always situated in the spongioplasmic walls of the primitive tubes.

centre of many large nerve-tubes. If we now examine similar nerve-tubes in transverse and longitudinal sections we will find that, their contents consist of a bundle of slender cylindrical primitive tubes, quite in the same way as that just mentioned. The only difference is that, their central primitive-tubes have a smaller diameter and thicker walls, and are more deeply stained than the peripheral ones, it seems, indeed, as if they may have been pressed more tightly together and thus been obliged to occupy a smaller space, as they have on the other hand got a firmer consistency, with thicker walls. In this axis, and in its neighbourhood, larger dark granules, as mentioned above, also occur more frequently than anywhere else, which perhaps contributes somewhat to the darker staining. In some nerve-tubes this concentration and forming of an axis is so far developed that it, in transverse sections (fig. 3), appears as a, by osmic-acid and hæmatoxylin, deeply stained spot in the centre of the tube. In this spot it is not easy to distinguish any structure or primitive tubes, but in extremely thin sections, and by very high powers, it is, however possible to see slender tubes with thick, deeply stained, walls or membranes in which granules occur. In longitudinal sections a longitudinal striation appears as in fig. 4. In one end (fig. 4, a) this section has passed through the periphery of the axis and, here, a striation is distinctly visible; in the other end (fig. 4, a') the section has, however, passed more through the centre of the axis and, here, the staining is so deep that almost no striation is visible, all the less from the section being somewhat thick.

Usually however this concentration is not so distinctly developed as here. Nay, we can indeed find every degree of development, from the primary state where no concentration at all can possibly be traced to have taken place, up to an axis as described. Fig. 2, t, t', t'' thus represent sections of fibres with different degrees of concentration, from a very slight one where we can only see that the primitive tubes are somewhat deeper stained in this central part than in the rest of the tube and in some nervn-tubes even this is not visible (t').

It is the nerve-tubes with such an axis that, previous writers have called myeloid fibres; with what right we will, on a later occasion, have an opportunity of examining.

However we consider this formation of an axis; in one thing we cannot be in doubt, viz., that the whole contents of the nerve-tubes, wether they have an axis or not, is of real *nervous* nature, because the constructing element in *the whole contents* throughout is, as we have seen, *the "primitive tube"*; that is the thread o which the whole rope is woven.

In transverse sections of some nerve-tubes, largish vacuoli or, as it were, primitive tubes, can be seen (fig. 5, t', t'', t'''; fig. 2, d). These vacuoli often occur in a peripheral layer just inside the sheath of the nerve-tubes (fig. 5, t', t''). I suppose, however, this appearance to be to a great extent artificially produced. In some cases the whole contents of the nerve-tubes has the same appearance (fig. 5, t_1) and looks as if it were principally constituted of large primitive tubes, amongst which only some few primitive tubes of the common size are seen. I do not feel disposed to suppose this appearance to be only an artificial and postmortem product, though I cannot with certainty account for its nature at present.

We have, hitherto, only mentioned nerve-tubes of a relatively large diameter; these are especially numerous in the longitudinal and oesophageal commissures, as will be seen from fig. 1, which represents a transverse section of an oesophageal commissure. We find, there, transsected nerve-tubes of very varying diameter.[1]) In a great many large tubes (t, t', t_1) a deeply stained axis is visible; in other large tubes (t'') no axis is visible. *n.t* represent somewhat smaller nerve-tubes running in a thick bundle along the centre of the commissure. *s.nt* represent very small nerve-tubes situated more peripherically.

Large nerve-tubes, usually having an axis, occur very frequently also in the peripheral nerves, vide fig. 7, t; they have generally, as will be seen, very stout sheaths, and are prominent.

On the other hand there are, especially in the peripheral nerves, a great multitude of extremely slender nerve-tubes; indeed, the peripheral nerves principally consist of such nerve-tubes (fig. 7, *nt*).

These nerve-tubes have so small a diameter that, I have, usually, only observed some few primitive tubes inside their sheaths (vide fig. 8, *nt*); in a large number of them I have even detected no primitive tubes, and I believe they are partly constructed of only one primitive tube, the sheath of which is, however, much stouter than the spongioplasmic sheaths inside the larger nerve-tubes.[2])

These slender nerve-tubes are, in most nerves, usually arranged or united in bundles, and are enclosed in neurilem-sheaths, larger fagots of these primitive bundles are again enclosed in larger and

---

[1]) We can find nerve-tubes with a diameter of more than *0.140 Mm.*; and we can find nerve-tubes with a diameters of less than *0.003 Mm.*

[2]) The diameter of these smallest nerve-tubes measures about .0017 Mm.

stouter neurilem-sheaths, and so on, concentrically, untill at last the whole nerve is enclosed in one *common external neurilem-sheath*.[1]) It is, indeed, very difficult to tell here where the separation of individual nerve-tubes really begins, if we do not, as I have done, consider the smallest tubes with distinct stout sheaths as representing, each of them, a nerve-tube, consisting of some few primitive tubes only, or, in a great many cases, of one primitive tube only.

When we examine a transverse section of a nerve, we will, as a rule, observe a great many dark granules or dots; these dots, appear, however, on closer examination, to be thickenings in the tube-sheaths along the longitudinal edges where several tubes meet (vide fig. 8, *p*).

The sheaths of the nerve-tubes. — Our attention has hitherto been directed to the contents of the nerve-tubes only. We will, now, before we leave the nerve-tubes of the lobster, pay a little attention to the sheath which envelopes the contents. This sheath consists of a connective substance which is, in my opinion, the same substance as the *neuroglia*, or »connnctive-tissue«, as many authors call it, which extends through the whole central nervous system of every animal I have had under investigation. This sheath of connective substance, or for the sake of brevity we will call it *neuroglia-sheath*, has, also, a great resemblance to the *spongioplasm* separating the primitive tubes, and it is, in fact, very often, extremely difficult to distinguish the two substances from each other viz. the *spongioplasm* occurring inside the nerve-tubes and the ganglion cells — as will subsequently be described — and the *neuroglia* enveloping the nerve-tubes and the ganglion cells with sheaths or membranes; the two substances are often so intimately united that it is really impossible to decide where the line of demarcation can be drawn. There is one difference, however, viz. that in the spongioplasm no nuclei occur, whilst nuclei occur in the neuroglia. LEYDIG's view that it is the same substance, spongioplasm, which penetrates from whithout into the nerve-system and even into the nervous elements, is, in my opinion, not yet sufficiently well substantiated. If I had to choose, I would, however, much prefer that theory to VEJDOVSKÝ's, according to which the inner connective substance (i. e. neuroglia) is a product originating in the ganglion cells; I really do not understand how, in that case, to account for

---

[1]) This separation into bundles and fagots is, however, les prominent or not present at all near the origin of the nerves in the central nerve-system.

the neuroglia-cells and nuclei, and how they can be produced from ganglion cells.

In my opinion there can be no doubt, but that the neuroglia is a separate tissue composed of cells springing from the ectoderm, just as the ganglion cells spring from other ectodermal cells.

The doubly marked outlines which are visible in nerve-tubes isolated in the fresh state, but which are specially distinct in macerated nerve-tubes, are produced by the sheath which, as a cylinder, envelopes every nerve-tube and forms its outer isolating layer. It consists of one or several concentric layers of connective substance or, as we just above called it, *neuroglia*, and is easily seen in transverse as well as in longitudinal sections; the sheaths of large nerve-tubes are especially very prominent and stout. When there are several layers; which is generally the case in large nerve-tubes, especially those of the peripheral nerves; then the innermost layer is the strongest, most differentiated, and refractive one (vide fig. 15). These concentric layers of the tube-sheaths are seen, in fig. 7, round the large nerve-tubes (t), and in some peripheral nerves they are very prominent.

In the sheaths, nuclei occur. These nuclei are, as mentioned, quite identical with the usual nuclei of the neuroglia; they have an oblong form, with a granular appearance, and are usually situated on the outer side of the sheaths (vide fig. 2, k; fig. 5, k; fig. 8); they occur, however, also on the inside of the sheath, consequently in the tube itself (vide fig. 4, k; fig. 5, k'; fig. 13, k).

Nephrops norvegicus.

We obtain very similar results as to the structure of the nerve-tubes on examining *Nephrops norvegicus;* in this respect Nephrops so very much resembles Homarus that it is really unnecessary to give any special description of it. We can, in fact, observe, in the large nerve-tubes, the same tendency towards a concentration of a sort of axis in their centre, though it is not so prominent in Nephrops as in Homarus, and neither have I observed so narrow and deeply stained axes in the former as I have in the latter.

In the transverse sections of many nerve-tubes, I have observed similar large meshes as those. I have mentioned in the nerve-tubes of Homarus (cmfr. fig. 5, t_1). They had often a very regular appearance and looked as if they really were large transsected primitive tubes.

In the Annelids we find a structure of the nerve-tubes very similar to what is described in Homarus.

Polychætes.

As representative of the *Polychætes* I have examined several species of *Nereis* (N. virens, N. pelagica and others) and also, occasionally, *Nephtys, Leanira* etc. and can only say that, in them all, I have found the same structure of the nerve-tubes repeating itself.

In the *Nereidæ*, to which I have especially paid attention, we find nerve-tubes of every thickness, ranging from the two gigantic nerve-tubes running along the centre of the ventral nerve-cord (vide fig. 14 and fig. 10) down to nerve-tubes consisting of only one primitive tube, like what is described in Homarus.[1]

The minute structure is, of course, easiest to study in the larger nerve-tubes. If we examine them under high powers of the microscope, in good preparations,[2] we find that quite similar *primitive tubes* constitute their contents, in the same way as described in respect of the nerve-tubes of Homarus. The differentiation in primitive tubes is, however, not so distinctly marked nor so easily seen as it is in Homarus, perhaps, partly, because the primitive tubes have a smaller diameter,[3] partly, because their spongioplasmic sheaths or membranes are thinner and less distinct. In fact, it often requires the best lenses, and very carefully prepared preparations, to see any structure in the contents, and this is evidently the reason why so many authors have not understood these structures, and have supposed these large nervetubes to be lymphoid vessels etc. Of a concentration towards an axis, there are only very slight indications in some rather slender nerve-tubes. The large nerve-tubes (»gigantic fibres«) have always primitive tubes, only slightly marked, and of the same size and appearance throughout. Fig. 10 represents the central part af a transverse section through the ventral nerve-cord, where one of the two central largest nerve-tubes (t) (of the other only a small part is seen), also some more ventrally situated, not so large nerve-tubes (t_1, t_2, t_3) and a great deal of the surrounding small nerve-tubes (nt) are seen. Fig. 9, b represents a longitudinal section of one of the largish nerve-tubes (cmfr. fig. 10, t_2);

---

[1] In my preparations I have found nerve-tubes with diameters measuring from .030 Mm. down to .0018 Mm.

[2] For this purpose the same method of fixing etc. can be recommended as we have recommended for the neve-tubes of Homarus (vide p. 85).

[3] In my preparations I have generally found the diameter of the primitive tubes to be about .0012 Mm. or less.

nt represents smaller surrounding nerve-tubes. As will be seen, there is, in these tubes, a longitudinal striation quite similar to what we already have experienced in Homarus.

The nerve-tubes of the peripheral nerves have quite a similar structure to those of the ventral nerve-cord.

The sheaths of the nerve-tubes are formed by the neuroglia, in the same way as they are in Homarus (cmfr. fig. 10); they are, however, not so stout, and nuclei occur very sparingly in them.

Lumbricus agricola.

In Lumbricus, the nerve-tubes in the ventral nerve-cord have a more uniform and relatively smaller size (the three »gigantic nerve-tubes« excepted [fig. 11, t, t, t_1]) than they have in the ventral nerve-cord of the Polychætes examined; the nerve-tubes have, however, a quite similar structure, their contents being composed of the same elementary constituents, viz. the primitive tubes, which certainly have a small diameter, but are, in the small nerve-tubes, only present in limited number; in some cases nerve-tubes consist of apparently only one primitive tube; it is consequently somewhat similar to what is described in Nereis and Crustaceans.

The nerve-tubes have neuroglia-sheaths which are very prominent in osmic-hæmatoxylin preparations, vide fig. 11, *nt*.

The nerve-tubes of the peripheral nerves have a quite similar structure to those of the ventral nerve-cord.

In the ventral nerve-cord of Lumbricus, there are, as is well known, three large tubes running, dorsally, through the whole length of the nerve-cord. These tubes, which are generally called »CLAPARÈDE's gigantic nerve-fibres«, have given rise to much dispute and in the views of the various writers very different functions have been attributed to them. Some writers have called them supporting rods, the function of which is to give support and rigidity to the ventral nerve-cord under the many movements of the animal; that is, for instance, VEJDOVSKÝ's view; other writers, again, consider them as homologous with the corda spinalis of the vertebrates; some writers call them vessels, etc. etc., and finally a great many writers call them nerve-tubes. Lately, LEYDIG has published a paper on this subject (l. c. 1886) in which he very emphatically maintains their exclusively nervous nature.

After having examined their structure, I do not think there can be any doubt that LEYDIG is right, and that VEJDOVSKÝ, who expresses himself with so much self-confidence, is entirely wrong. It was indeed to be hoped that those organs might at last enjoy that rest wich they so well deserved and that they might now be left in peace in the rubric of nervous elements.

Seeing that, comparatively, few authors have been in doubt as to the nervous nature of the large nerve-tubes of the Polychætes, it is really very strange that there has been so much dispute about these organs in Lumbricus. — If we examine them closely under high powers of the microscope, in carefully prepared sections, we will, instead of the homogeneous contents usually described in them, find a contents with quite the same structure as above described in the large nerve-tubes of Nereis, it being composed of a large bundle of *primitive tubes*, the spongioplasmic sheaths of which are, however, very thin and but little differentiated; the primitive tubes are thus extremely difficult to observe, and this is naturally the reason why no author has noticed any striation in these large tubes, and why their contents has always been described as being homogeneous.

I have not been able to find any spongioplasmic reticulation or septa, similar to what LEYDIG has described and illustrated (l. c. 1886 p. 594). I cannot therefore explain their absence in any other way, than that his preparations have not been quite succesful, there may, perhaps, have been some irregularities in them produced by shrinking of the tube-contents, which indeed very often happens, especially in those thick tubes. From my own sad experience, I can testify that it sometimes happens in spite of an apparently very careful preparation. Judging from LEYDIG's illustration I should also say that such has been the case.

The whole contents of the three nerve-tubes consist of primitive tubes, having the same size and diameter throughout, as will be seen in fig. 11, and no concentration towards an axis is visible in the centre of them.[1])

The sheaths of the nerve-tubes. — The three large nerve-tubes are surrounded by very thick and prominent neuroglia-sheaths consisting of many layers or membranes of connective substance. When LEYDIG says that they are closely surrounded by several slender nerve-tubes, he is scarcely quite correct. I have certainly observed nerve-tubes between them, and in the neuroglia surrounding them, but they are scarce, and, in my opinion, the three large nerve-tubes are principally surrounded by connective substance, or neuroglia, forming thick sheaths round them. This neuroglia does not, however, differ from the neuroglia of the rest of the nervous system, in anything, else than that it occurs in thicker layers than is

---

[1]) In my preparations the diameter of the primitive tube measured about .0016 Mm.

generally the case, and is more sparingly mixed with nervous elements. As will be seen in fig. 11, it has, also, an appearance somewhat different from the rest of the fibrillar mass, exhibiting, as it does, a prominent disposition to form concentric layers or membranes round the tubes. Between the tubes may also be seen septa (fig. 11, *s*), apparently of the same substance, issuing from, or rather adhering to, the *neurilem-sheath* or *perineurium*, enveloping the ventral nerve-cord inside the muscular layer (fig. 11, *m*). In the neuroglia-mass surrounding the three large nerve-tubes nuclei occur very often (fig. 11, *k*), they are oblong, have a granular appearance and are quite similar to the common neuroglia-nuclei.

The Mollusca.

Judging from the latest descriptions of the nervous system of the Molluscs, we should expect, here, to find very extraordinary conditions. HALLER assures us that there exists no connective-tissue within the nervous system of the more primary Molluscs (Chiton, Patella, Rhipidoglossa etc.) and that there really exists *no* nerve-fibres or, as I call it, *nerve-tubes* only »Primitivfibrillen« which unite in large bundles to form the peripheral nerves. The same statements are quite recently made by RAWITZ (as mentioned p. 70) as to the nervous system of the Acephales; this latter writer seems, however, in these as well as other respects, to walk very closely in the footsteps of HALLER.

It was therefore with no small interest that I began the investigation of this group. I chose *Patella vulgata* for my purpose, because it was on the one hand a large species, and a rather primary mollusc, and on the other hand I could get plenty of it, here, in Bergen.

I must confess, I had some doubt as to the correctness of BÉLA HALLER's statement that no connective-tissue (i. e. neuroglia) existed, and my researches have not at all diminished my doubts; on the contrary, I found a well developed neuroglia, with neuroglia-nuclei of the typical form, and sheaths enclosing nerve-tubes. This neuroglia is, however, not so well developed as we have seen it in Annelids and Crustaceans. It took me some time, therefore, to come to a full understanding of the real structure of the nerves and the nerve-tubes. Fresh isolated nerves, as well as nerves macerated in different ways, I could, usually, though always with some difficulty, split up into fibrillæ, which, however, varied somewhat in size (fig. 17). In sections, longitudinal as well as transverse, I came to similar results, and could really distinguish longitudinally running darker stained fibrillæ, of somewhat varying size; in transverse sections they appeared as dark points; in longitudinal srctions as dark lines, and in oblique sections I could see them as short semi-erect rods. There could be no doubt that real fibrillæ of some kind were present, and I was almost disposed to consider HALLER's description of the nerves, as consisting of bundles of »Primitivfibrillen«, to be right. Besides these fibrillæ I could, however, in transverse sections see a reticulation, as if produced by transsected tubes; in this reticulation the transsected fibrillæ were always situated in the walls of the meshes, and especially in the junctions of the walls of several meshes (i. e. nerve-tubes) the whole had, consequently, somewhat of a resemblance to transverse sections of nerves of Annelids and Crustaceans. I could, however, in preparations obtained in my usual ways come to no certain results. There was, besides, also another circumstance which made me certain that those fibrillæ could not be real nervous fibrillæ as HALLER and RAWITZ suppose, this was their staining, which was quite that specially pertaining to neuroglia.

Finally, I succeeded in finding a method of preparing the nerves which gave a quite clear idea of their structure. This method consists in treating pieces of Patella (the pedal muscular disc in which the pedal nervercords are imbedded) for about 48 hours in osmic acid (1 %) then washing them, afterwards cutting them and staining the sections as described above (p. 77).

Examination of sections obtained in this way left no doubt as to the structure of the nerves; they contain nerve-tubes having distinct neuroglia-sheaths, in which thicker longitudinally running fibres are situated (cmfr. fig. 19).[1]) These neuroglia-fibres are the »Primitivfibrillen« of HALLER and RAWITZ. Besides this, there occur stouter neuroglia septa orriginating in the outer neurilem-sheath. They ramify and divide the nerve-tubes into larger or smaller bundles (fig. 19, s, s'). Fig. 19, k, k are nuclei occurring in these neuroglia-septa.

The nerve-tubes are of somewhat varying size, and are usually slender and seem to contain but few *primitive tubes*. I measured nerve-tubes having a diameter of .002—.006 Mm. In some nerves the nerve-tubes are, however, much smaller. Fig. 20 is, for instance,

---

[1]) A glance at the origin of the small lateral nerves a' and b' (fig. 19) gives a very clear idea of the structure of the nerve-tubes. s', s' are the neuroglia-septa which separate the bundles of nerve-tubes, forming these small nerves, from the rest of the great nerve. In a and b some of these nerve-tubes are transversally transsected in a' and b' they are longitudinally transsected. It is very distinctly see that the »fibrillæ« in fact belong to the tube-sheaths.

an illustration of a longitudinally transsected nerve, drawn under higher powers than fig. 19. The nerve-tubes are, however, narrower in fig. 20, than they are in fig. 19 (cmfr. a' and b').[1]

The reason why the longitudinal neuroglia fibres in the nerves and pedal nerve-cords are isolated in macerated preparations is, evidently, that they are stouter and stronger than the rest of the tube-sheaths, and thus are separated from them. In carefully treated macerated preparations it is, however, possible to see the nerve-tubes. Fig. 17 and 18 represent, for instance, such preparations, where nerve-tubes with a longitudinal striation are distinctly visible, and even to some extent isolated; an isolation of them is, however, as a rule, extremely difficult, for the reason just indicated above. In the extremities o these nerve-tubes we can see, in some of them, isolated fibrillæ, which, partly, are neuroglia fibrillæ, partly, spongioplasmic fibrillæ. Pearls of hyaloplasm are adherent to the sides of them (fig. 17, a, b, d).

The Ascidians.

Finally, I will here mention a group of invertebrates which some time ago were the subject of my study, the results of which I have hitherto only given in a preliminary report (l. c. 1886). This group is the *Ascidians*. The peripheral nerves of the Ascidians have a structure very similar to what is described of the peripheral nerves of Homarus (cmfr. fig. 7), exept as regards the large tubes; those do not occur in the Ascidian-nerves.

The whole nerve is, in the Ascidians, divided by the neuroglia, or inner neurilem, into large bundles, these are again divided into smaller and these, again, into still smaller bundles, the subdivision being repeated until we at last arrive at the nerve-tubes, which are very slender, and contain but few primitive tubes; a great many nerve-tubes seem to consist, even, of only one primitive tube, something similar to what we have described in Homarus. Very often, it is even very difficult to decide what are only primitive tubes, and what are nerve-tubes, and again what are bundles of nerve-tubes, as there is often seen, in transverse sections, a subdivision into smaller and smaller tubes, and, the higher and higher we employ the powers of the microscope the more and more do we trace out minute tubes. At last, however, under very high powers we are able to observe

---

[1] Sometimes large vacuoles are seen in transverse section of nerves (cmfr. fig. 19, c, d); whether these vacuoles really are transsected nerve-tubes I am not in a position to decide at present.

some very slender tubes inside which no subdivision is visible, and these we must supose to be the primitive tubes, a small bundle of which generally forms a nerve-tube.

Fig. 21, which represents the transverse section of an anterior nerve of *Phallusia venosa*, will, I hope, give easily, at a glance, a correct idea of the structure of the peripheral nerves of the Ascidians.[1]) Fig. 22 represents a part of a longitudinal section through a nerve, seen under high powers of the microscope; the primitive tubes are, here, visible, being transsected in their curved course, the curvature in which arises from the contraction of the nerve. *ts, ts* are the sheaths of the nerve-tubes. Some of them (*a* and *b*) are stouter and more prominent than the others, those are probably transsected neuroglia-septa enclosing bundles of nerve-tubes. In one of these septa is seen a neuroglia-nucleus (*n*).

The division of the nerve-tubes into different smaller and larger bundles is, in the Ascidians, as well as in the Molluscs, Crustaceans etc. less prominent near the origin of the nerves; close to the nerve-roots there is, in the Ascidians, scarcely any division visible, and a great many of the nerve-tubes appear, even, to be broken up into nerve-tubes of a much smaller diameter; the reason of this appearance we will in a later chapter have an opportunity of discussing.

The structure of the nerves of the Ascidians reminds much of what we have found in the Molluscs. The nerve-tubes are separated by only very thin sheaths (cmfr fig. 22, *ts*) in which thickish fibres are situated, especially along the concreting longitudinal edges; in transverse sections these fibres, when transsected, appear as darker dots, situated in the corners of the meshes, and are produced by the transsected sheaths of the nerve-tubes (vide fig. 21).

Summary.

The results of, these, our researches on the minute structure of the nerve-tubes of various invertebrates, we may, I think, assume to be applicable to the nerve-tubes of all invertebrates of such a high organisation as to posess a well developed nervous system, i. e. all *invertebrated bilaterates*. From LANG's memoirs, we may gather that the Polyclades and Trematodes have nerve-tubes of the same type described in this chapter (cmfr. p. 52—54); from FRAIPONT's

---

[1]) The primitive tubes are not seen in this illustration.

memoir on the Archiannelids we may gather that this is the case also in those animals (cmfr. p. 57), etc. etc.[1]

Of the results of our researches we may therefore give the following summary:

1) The nerve-tubes of the invertebrated bilaterates consist of *an external consistent sheath* with a viscous *contents*.

2) *The sheaths* are formed by, or belong to, the connective substance extending through the whole nervous system, and which I call neuroglia.[2]) In these sheaths nuclei (neuroglia-nuclei), occur more or less sparingly.

3) The contents of the nerve-tubes consist of *primitive tubes*, which are extremely slender tubes or cylinders, separated from each other or rather formed by membranes or sheaths of a firm supporting substance, *spongioplasm*, very much resembling the neuroglia-substance; these slender cylinders of spongioplasm contain a hyaline, viscous substance, *hyaloplasm*, which is the real nervous substance, and which very often is exuded from fresh isolated nerve-tubes in form of small hyaline pearls.

Fibrillæ and fibres, as most writers describe them, do not, consequently, in my opinion, exist.

4) A concentration towards a kind of axis is visible in a great many largish nerve-tubes of *Homarus* and *Nephrops*; this axis may be more or less narrow, and consists of a bundle of central primitive tubes which have stouter spongioplasmic sheaths and smaller diameters than the other primitive tubes. A similar concentration in the centre of the nerve-tubes can, as a rule, not be observed in the other animals examined; only a slight indication of it, I believe to have observed in some nerve-tubes of Nereis.

2. The structure of the ganglion cells, and their processes, in invertebrates.

Homarus vulgaris

The contents of the ganglion cells of the lobster consist of nucleus and protoplasm enclosed in a thicker or thinner membrane or sheath.

---

[1]) In Myzostoma I have previously indicated a similar structure.

[2]) In my memoir on *Myzostoma* (1885) I have called it »the inner neurilem«, as mentioned before.

These membranes or sheaths of the ganglion cells have the same structure as the sheaths of the nerve-tubes, and are formed of the same substance, viz. the *neuroglia;* nuclei, having the appearance of common neuroglia-nuclei, occur frequently in them, and are generally adherent to their outer side; I have seldom observed nuclei situated on the inner side as described of the nerve-tubes. The sheaths, enveloping the cells, very often consist of several layers or membranes, similar to what is described of the sheaths of the nerve-tubes.

I can not decide, whether inside these layers of neuroglia-substance there also occurs a thin membrane, spongioplasmic membrane (different from the neuroglia-sheath), belonging to and arising from the protoplasm of the ganglion cells. It would, consequently, be a »cell-membrane« as many writers describe it to be, but the existence of which other writers deny.[1] The importance of such a membrane existing, or not existing, is, however, in my opinion, not great as may be seen from the subsequent description of the structure of the cell-protoplasm.

When several writers deny the existence of a cell-membrane, as well as a membrane of connective substance, I think that must arise from their examining macerated preparations, and imperfectly stained sections. In macerated preparations, it is often very difficult to distinguish the enveloping membrane from the cell-protoplasm, it having a concentric striation which is very similar to the outer layers of the protoplasm; neuroglia-nuclei are, however, usually adherent to its outer side and indicate its real neuroglia nature (cmfr. fig. 23, *n*).

In successfully stained sections, e. g., sections stained with HEIDENHAIN's hæmatoxylin-method (vide p. 76), the neuroglia membranes are, even under the lower powers of the microscope, distinctly visible (fig. 24—29). In such preparations, these membranes also appear to be intimately connected with the neuroglia-reticulation extending between the ganglion cells (cmfr. fig. 24), and they look as if they belonged to that substance, which, in my opinion, is also

---

[1] If such a special spongioplasmic membrane occurs, which is, in my opinion, not improbable, it would, I think, be so intimately connected with the surrounding neuroglia-membranes (cmfr. my subsequent description of the cell-protoplasm and its reticulation) and have an appearance so perfectly similar to them that, I cannot understand how authors, working with macerated preparations, can describe special cell-membranes as being different from the surrounding »connective-tissue« membranes.

the case, as will also be subsequently referred to. In some parts of the ganglia, where small ganglion cells are situated closely together, the neuroglia-membranes form cavities in a similar way as the waxen-walls in a honeycomb, and one ganglion cell is situated in each cavity (vide fig. 40).

The membranes also extend into, and envelope, the processes of the ganglion cells (vide fig. 26). Round the processes of a great many cells, sheaths of a quite peculiar structure occur, these sheaths will be mentioned in the description of the processes.

The protoplasm of the ganglion cells of the lobster has been described in different ways by the many authors who have treated the subject. In one of the later and more important memoirs on this subject, viz. that by FREUD (1882), it is, as we have seen (p. 34), described as consisting of two substances, a reticulated substance, and a hyaline viscous substance.

FREUD has specially examined ganglion cells in the fresh state. I have also examined them in the fresh state, but could not arrive at a lucid conclusion as to the structure of their protoplasm, neither could cells isolated by maceration afford much better information in this respect.

I saw a kind of reticulation, as described by FREUD. This reticular appearance was especially distinct in macerated preparations (cmfr. fig. 23) the meshes having a circular shape (they have become too distinct in fig. 23; the illustration not being a successfull one); I could, however, not convince myself whether this appearance was really produced by a substance having a reticular structure or not. Besides this reticulation I could generally to some extent trace out a concentric striation (cmfr. fig. 23, it was, however, never so distinct as it appears in the illustration).

On examination of carefully prepared sections[1]) somewhat more light was, however, thrown on the subject, though I must confess, that there are several points in which I could still wish to come to more clearness.

Fig. 24 is a part of a section through two large ganglion cells in an abdominal ganglion. We can there see that the membranes *cm*) enclosing the cells are stout, and are very distinctly marked.

---

[1]) The best preparation-method for this purpose I found to be, fixing in *chromo-aceto-osmic acid* and staining with Heidenhain's hæmatoxylin (cmfr. p. 76), or fixing in Lang's sublimate-solution and staining in Heidenhain's hæmatoxylin (cmfr. p. 77). To give the preparations, afterwards, a nuclear staining by Delafield's hæmatoxylin is a very good practice (cmfr. p. 76).

In connection with these membranes we find, however, a very strange structure. In B it has the appearance as if a series of deeply stained oblong corpuscles were situated along the inner side of the membrane. Upon closer examination we will see that these corpuscles are, as a rule, connected with the membrane by very slender filaments. In some cases these filaments are even very thick and the dark corpuscles appear thus to be transsected fibres or septa issuing from the membrane (cmfr. fig. 24, *b*). On the other side we will, on close examination, find slender filaments issuing from the inner extremities of those dark corpuscles or fibres and penetrating into the protoplasm of the cells; indeed, we will see that they anastomose and form a reticulation, extending through the cell-protoplasm and distinctly visible in its outer layers. The meshes of this reticulation are also largest in the outer layers near the surrounding membrane (vide fig. 24). They have, here, very often, an oblong form, going in a radiate direction towards the centre of the cells, and having, sometimes, the appearance as if the meshes were formed by filaments radiating from this centre to the surrounding cell-sheaths; in fig. 24 B such an arrangement is very distinct. Within these large meshes smaller meshes are generally seen (fig. 24). These smaller meshes have, however, a somewhat different appearance, they being not so distinctly visible; I am not sure whether they are formed in quite the same way.

Towards the inner part of the cell-protoplasm (which is more deeply stained) the meshes of the first mentioned reticulation become much smaller and become similar to the small meshes just mentioned, indeed, they cannot be distinguished from each other. The meshes are so narrow that they are visible only under the higher powers of the microscope, and, even then, not very distinctly; in small cells they are especially very difficult to observe.

In the filaments of this reticulation, granules and thickenings occur, giving the cell-protoplasm, seen under lower powers of the microscope, a granular appearance. These thickenings are, especially in the large ganglion cells, very prominent in the outer layers of the protoplasm.

In the small meshes of this reticulation, a hyaline substance is suspended, very similar to the hyaloplasm of the nerve-tubes.

The question which now, as regards the reticulation, very naturally forces itself upon us is, whether is it a real spongy reticulation extending through the protoplasm of the cells, or is it a reticulation produced by a transsection of tubes in the same way as we have seen in the nerve-tubes. My opinion upon this point is, that this reticulation is, to some extent, a real one having partly a quite spongy structure without the formation of tubes; to some extent the latter must, however, also be the case, as we shall soon see. The substance forming the reticulation has, as mentioned, a great resemblance to the sheaths of the cells as regards its staining etc., and it is, in my opinion, the same substance which forms the sheaths of the primitive tubes in the nerve-tubes, viz. *spongioplasm*.

Before we subject the significance of this substance to a further examination I think it will be well to look somewhat at the structure of the processes of the ganglion cells and their origin in the cell-protoplasm.

The contents of the processes, within the cell, its structure and origin in the cell-protoplasm. — The ganglion cells of the lobster are, as a rule, *unipolar*; if they have several processes, there is always one of the process which has a special appearance, and which is the real *nervous process*; if other processes are present they are, as I call them, *protoplasmic processes*, and they have, in my opinion, a nutritive function, as will be subsequently mentioned. At present we will only consider the structure and origin of the nervous process.[1]

In well prepared sections, a distinct longitudinal striation of these processes is easily observed. On observation of transversally transsected processes, it is seen that this striation is produced by primitive tubes having the same structure, and constituting the nervous processes in the same way, as described of the nerve-tubes.

As regards the staining, the contents of the nervous processes differs from the protoplasm of the ganglion cells; the latter having generally a much darker staining. This difference is, for instance, very prominent in fig. 37 & 38, where the contents of the longitudinally transsected processes have a very light staining and a distinct striation.

The origin of the nervous processes differs somewhat in the various cells. In fig. 26 it has the appearance, as if the contents of the nervous process arises from a convergence of primitive tubes from the whole body of cell-protoplasm, in which they have a uniform extension, and to which they generally give, to some extent, a con-

---

[1] I have in reality, never observed a multipolar ganglion cell of quite indubitable multipolar shape, in the nervous system of the lobster. I am afraid that it has very often been neuroglia-fibres issuing from neuroglia-sheaths which have given rise to descriptions of multipolar ganglion cells.

centric striation. Some of the primitive tubes we can in many preparations even trace, for some distance, in their course through the cell-protoplasm, of which they are not, as we have previously seen (cmfr. reticulation described p. 101), the only constituent, but of which they are, however, a principal part.[1]) These primitive tubes have the same structure and diameter as those we have described in the nerve-tubes; they consist of hyaline contents, *hyaloplasm*, enveloped in sheaths of *spongioplasm*, which has the same staining and appearance, and is evidently the same substance, as that previously described (vide p. 101), as forming the reticulation in the cell-protoplasm. In fig. 27 and 28 it is very distinctly seen that these spongioplasmic sheaths of the primitive tubes are intimately connected with the peculiar peripherically situated fibres (fig. 27, *b*; fig. 28, *b*) issuing from the surrounding neuroglia-membrane.

In a great many ganglion cells these primitive tubes have not, however, such an uniform extension through the protoplasm as illustrated in fig. 26. In the large cells they are generally united to bundles, distinctly distinguished from the rest of the protoplasm. In successfully stained sections,[2]) where they are transversally transsected, they are distinctly visible as larger or smaller light areas situated in the deeply stained protoplasm. Very often, when the bundles of primitive tubes circulating in the cell-protoplasm are numerous and small, we get sections having the appearance illustrated in fig. 25 and 27.

In ganglion cells containing such bundles of primitive tubes circulating in the protoplasm, the nervous process arises in such manner that the bundles unite to form the process, as illustrated in fig. 27 and 25. This union takes place generally within the cell-protoplasm, and the contents of the nervous process has then, for some distance, an undivided course through the cell-protoplasm, and can be traced as a large light area through a series of transverse sections (cmfr. fig. 28 and 29). It is surrounded by thicker or thinner deeply stained fibres (fig. 28, *s* and *b*; fig. 29, *s* and *s'*) of the same substance as forms the reticulation of the protoplasm (cmfr. p. 101). Some of these

---

[1]) That it really is primitive tubes, with slightly stained hyaline contents, and not »fibrillæ« which circulate in the protoplasm of the ganglion cells, is easily seen in preparations deeply stained by hæmatoxylin (e. g. fig. 37); they are distinctly visible as light concentric lines where they are longitudinally transsected (cmfr. also fig. 38).

[2]) The fixing method and staining described p. 76 and p. 85 can especially be recommended for this purpose.

fibres are, in some sections, seen issuing from the sheath of the ganglion cell. Similar fibres can also be traced into the prosesses (fig. 27, b) where they have a longitudinal course (fig. 30, sf', sf''). Sometimes, in the cells, they are so closely situated that it looks as if they almost formed a sheath (cmfr. fig. 28) surrounding the process-contents for some distance into the cell-protoplasm, where they gradually disappear and obtain the same appearance as the reticulated spongioplasm, which spongioplasm often apparently gives, even the smaller bundles of primitive tubes in the protoplasm, a kind of relatively firmer surrounding layer resembling a thin membrane, which layer is generally more deeply stained by hæmatoxylin than the surrounding protoplasm (vide fig. 25, 27, 28, v and 29, v, v').

Sometimes the primitive tubes are united to a few and very large masses situated peripherically in the cell.[1]) Fig. 38, A represents a section through such a cell; v and v' are the transsected masses of primitive tubes. In fig. 39, which represents a more laterally running section through the same cell, it is seen what an extensive distribution these masses can have in the peripheral parts of the cell, the section passes almost exclusively through such a mass of primitive tubes. If we examine these light areas under high powers of the microscope, we see that they exhibit the same reticulation which we know from transversally transsected nerve-tubes and nervous processes, and which reticulation evidently arises from the transsection of primitive tubes.

In many ganglion cells we find similar peripheral masses, which are not, however, so distinctly defined as those just described. We can, indeed, find every stage of transition, from cells such as those illustrated in fig. 38, A to cells with a diffusive, uniform extension of a lightly stained mass in their whole peripheral layers, such cells are, for instance, illustrated in fig. 38, B and fig. 37. These peripheral layers, of lightly stained mass, can have a more or less distinct demarcation from the mesial, deeply stained, part of the protoplasm, or they can also have a quite successive transition into it.

In the protoplasm of most ganglion cells we will, indeed, find a

---

[1]) In macerated preparations these masses are often seen, having the appearance of vacuoli situated especially in the peripheral layers of the protoplasm of the cells and along their margins. For a long time I believed this appearance of the cells to be of artificial nature, produced by the macerating agents. It was first on examination of successfully prepared sections that I learnt their real nature. In macerated cells, the striated process-contents could sometimes be traced into the cell-protoplasm, it having, for some distance, an undivided course (vide fig. 23, A).

tendency to stain most deeply in their mesial part, whilst the peripheral layers are generally very light (cmfr. fig.s 24, 26, 27, 37, 38 B, 40). The reason is probably, to some extent, that the primitive tubes occur in greater plenty in the peripheral layers, or rather that they are situated closer together, and that they are not to such extent separated by layers of the substance which occurs in plenty in the mesial parts of the protoplasm and give to it the deep staining (vide sequel). In these lightly stained layers, a spongioplasmic reticulation, described above (cmfr. fig. 24), is generally more or less visible, in fig. 37 it is, for instance, very distinct, and thick fibres also occur (vide also fig. 38, B).

In other cells such a reticulation is but slightly visible, and we find only circular meshes, which probably are transsected primitive tubes, this is especially the case in cells where these layers are confined to masses distinctly defined from the rest of the protoplasm as, for instance, in fig. 38, A. Here, only a few such fibres or membranes of spongioplasm are visible (s').[1]

In ganglion cells with such peripheral masses of primitive tubes, the nervous processes do not spring directly from these masses but seem to get their whole contents from the mesial deeply stained

---

[1] I think it must have been ganglion cells with similar light peripheral masses which *Freud* mentions l. c. 1881 p. 29—30. He says: »An den grossen unipolaren Zellen erscheint eine oft sehr breite, einen grösseren oder geringeren Theil der Zellperipherie einnehmende Zone, welche durchaus homogen und dem Kerninhalt ähnlich ist.« He believes »dass diese homogene Zone durch den als »Zwischensubstanz« beschriebenen Bestandtheil des Zellleibes gebildet wird, aus welcher die netzförmige, dunklere Substanz sich gegen den Kern zurückgezogen hat. Es finden sich auch häufig genug Zellen, an denen zwei homogene Randpartien durch einen dünnen Strang dunklerer, genetzter Substanz, welcher noch an der Peripherie festgehalten ist, getrennt werden.« Freud supposes, consequently, those »homogene Randpartien« to be appearances produced on the death of the cell. In the sympatic ganglion cells, he describes somewhat similar masses of hyaline substance, which he, however, rather believes to be a normal appearance belonging to the live-state of the cells.

It may here, also be mentioned that *Freud* describes (l. c. 1881 p. 26) and illustrates (fig. 1 & 5) nervous processes having a peculiar origin in the ganglion cells. He says of them: »In manchen Zellen ist ein Übergangsstück zwischen Zellleib und Nervenfaser nicht vorhanden; die Nervenfaser entspringt in anderer, sehr eigenthümlicher Weise. Dieselbe schmiegt sich nähmlich in Gestalt eines hellen Halbringes der Peripherie der Zelle an, um dann in's Innere des Zellleibes einzutreten.« *Krieger* (in his dissert. 1879) has before *Freud* (as Freud himself states) described similar appearances. I think there is no doubt that these structures are of the same nature as those described above respecting the origin of the contents of the nervous process (vide p. 103—104).

protoplasm. This is, for instance, very distinctly visible in fig. 37, as well as in fig. 38, A. I have examined a great many cells having quite similar appearance, but in them all I found the same relation between the protoplasm and the contents of the nervous process. Although the contents of the process is often surrounded by a lightly stained mass at its entrance into the cell, it can, however, be seen passing to the mesial deeply stained protoplasm (cmfr. fig. 37 and 38), without giving off tubes to the light mass, it having an undivided course for a shorter or longer distance into the protoplasm. In this course it is, as described above, distinctly visible, and is surrounded by thick spongioplasmic fibres (fig.s 37, *sf'*; fig. 38, *sf*). At some distance from the entrance, the primitive tubes constituting this process-contents spread, and are lost in the protoplasm of the ganglion cell; generally, I could, however, trace some of them for some distance in their course through this deeply stained protoplasm.[1]

To explain what relation there may be between the primitive tubes of the nervous process and the peripheral masses of primitive tubes in these cells is, of course, an extremely difficult matter. In fact, I have not been able to trace out the connection between them. We have, as mentioned, found every transition-form of ganglion cells, from cells in which the contents of the process is spread at once on its entrance into the cell (fig. 26), to cells where the contents of the process passes undivided, and distinctly marked for a longer or shorter distance into the protoplasm (fig.s 25, 28, 29, 37, 38) and afterwards is uniformly broken up and spread in the protoplasm (fig.s 37, 38) or is to a greater or smaller extent broken up into bundles of primitive tubes (fig.s 25, 27). We have also found every transition-form, from ganglion cells having a protoplasm which is uniformly deeply stained in the mesial part, and the staining of which gradually passes over into a lighter one towards the peripheral layers (fig. 26), to ganglion cells with a protoplasm which is deeply stained in the

---

[1]) Sections similar to those illustrated in fig. 37 or fig. 38, A, can certainly very easily give rise to a belief in nuclear processes. If one had a little predisposition to find nuclear processes, one could easily, for instance, in fig. 38, A combine the process with the nucleus, and suppose the reason why this connection was not seen in the section was, that the section had not been quite successful. I think, therefore, that the description of nuclear processes can to a great extent be ascribed to processes like those I have illustrated, which seen in isolated cells can of course, in a still higher degree, give the appearance of being connected with the nuclei. *Krieger* has already made the same supposition regarding those peculiar processes described by him, and which are, in my opinion, processes originating in somewhat similar way as here described (cmfr. fig. 23, A & fig. 25).

mesial part, but has distinctly defined light masses in its peripheral layers (fig. 38, A), or in which plenty of distinct, lightly stained bundles, or small masses, of primitive tubes occur (fig. 25).

There is evidently a great variety in the structure of the protoplasm, as well as in the origin of the nervous process in the ganglion cells. I do not think, however, that this difference of structure can have any deep physiological significance, because we find every stage of transition from one form to another. The peripheral constituent of the protoplasm in all cells, is evidently the primitive tube. The small circular meshes which are found in sections, every where in the protoplasm, in the deeply stained as well as in the lightly stained parts, are probably only transsected primitive tubes which in the deeply stained parts of the protoplasm are more separated from each other by thicker layers of a deeply stained substance than they are in the light parts.

Where, and how, the primitive tubes terminate, or rather how they originally are formed, I cannot say.

In a few cases I have, in the peripheral parts of some cells, believed to see slender tubes pass to the enveloping neuroglia-sheath. This appearance can, however, also be occasioned by the above mentioned spongioplasmic fibres issuing from the sheath. If such radiating tubes were really present in all ganglion cells, we could, perhaps, easily understand the statement of Dr. RAWITZ regarding the ganglion cells of the *Acephales;* viz. that hyaline small pearls were exuded from the cells on pressure. These pearls consist, I think, of hyaloplasm, and are probably exuded from primitive tubes, which must either terminate in the envelope of the cell or must have been broken by the pressure.

If we now consider what we have learnt of the protoplasm of the ganglion cells, we may sum it up in the following results:

The cell-protoplasm is composed of *primitive tubes* consisting of *hyaloplasmic contents* and *spongioplasmic envelope,* — further, of a *spongioplasmic reticulation,* extending through the protoplasm and also intimately connected with the surrounding cell-sheath — and, finally, of a *hyaline substance* having much resemblance to the hyaloplasm of the primitive tubes.

The primitive tubes circulate, as we have seen, in the protoplasm, giving it often a concentrically striated appearance (fig. 23). In large ganglion cells they are frequently united into a number of thicker or thinner bundles, sometimes these even grow to considerable masses, generally situated peripherically in the cells (cmfr. fig. 30).

If we try to explain the connection and relation of these constituents of the protoplasm to each other, we have indeed no easy task. And it is, especially of this, I could wish to have obtained more light.

That a principal part of the protoplasm of the ganglion cells consists of primitive tubes, I think must be evident from what is already described; that a great number of the small meshes, seen in sections of the protoplasm of the cells (in the peripheral part as well as in the mesial one) are transsected primitive tubes is, I think, very probable.

There must, however, in the protoplasm of the ganglion cells be something also present, besides primitive tubes, because the protoplasm is stained by hæmatoxylin and other agents in a manner very different from the contents of the nerve-tubes, which exclusively consists of primitive tubes. By hæmatoxylin, the protoplasm is generally stained very deeply, except those bundles or masses of primitive tubes which get a similar staining as the contents of the nerve-tubes. This deep staining may, certainly, to some extent be explained by a rich occurrence of spongioplasm (cmfr. the spongioplasmic reticulation described above p. 101). I feel, however, disposed to think, that besides this spongioplasm another (myeloid?) substance occurs, perhaps in connection with it. Spongioplasm with this fattish substance [1]) is, I suppose, situated in thicker or thinner layers between the primitive tubes running through the protoplasm, and separates each of them more widely than is the case in the nerve-tubes. Within the bundles or masses of primitive tubes, occurring in large ganglion cells, this substance is not present, nor is there apparently more spongioplasm than is generally present in the nerve-tubes or the nervous processes.

The structure of the nervous processes and their sheaths, in their course outside the ganglion cells. — The contents of the nervous processes is, as already mentioned, composed of primitive tubes, in the same way as the contents of the nerve-tubes. In its staining it is differentiated from the protoplasm of the

---

[1]) That it is a fattish (myeloid?) substance is, I think, probable from the dark or brownish staining of the cell-protoplasm by osmic acid. This is, however, also the case with the common spongioplasm.

ganglion cell, this difference being, as we have seen, already distinct within the cell itself. It has, in preparations often, the same light staining the whole way from the origin of the nervous process till its entrance into the dotted substance (fig. 35); generally it is, however, stained in a quite peculiar manner by hæmatoxylin. Fig. 30—33 represents various sections through a such process; it is the same process the contents of which is also seen in fig. 28 and 29. At the agress from the ganglion cell (fig. 30), the contents of the nervous process has the same light staining as already described. At some distance from the cell the contents begins, however, to get a darker, more black staining, especially in its peripheral layers (fig. 31). At the same time the diameter becomes also diminished. At a greater distance from the cell this blacker staining increases, and the whole contents is stained quite dark or black (fig. 32), the diameter of the process is also still more diminished. In fig. 33, the same process is seen divided into two branches (*a* and *b*) which have the same black staining as above. This black staining they retain till their entrance into the dotted substance. The branchlets issuing from them are also stained in the same manner. It is, of course, very difficult to say what is the reason of this dark staining. I do not think, however, that it is of qualitative nature. To some extent it may, perhaps, be explained by the relatively small thickness of the processes, and by their situation in a reticulated neuroglia through which the fixing as well as the staining agents easily find their way to these structures. It is, however, strange that, in the same section, one can find some processes with dark contents, and other processes with light contents, similar to the contents of the nerve-tubes (cmfr. fig. 35 & 36). The branchlets issuing from these processes have similar lightly stained contents, consisting of primitive tubes (cmfr. fig. 35, *br*; fig. 36, *br, br'*).

Spongioplasmic fibres similar to those surrounding the process-contents within the ganglion cell (cmfr. fig. 28 & 29, also, generally occur in the process itself. They have a longitudinal course along the process, inside the sheath (vide fig. 30, *sf'*, *sf''*) and can often be traced to a considerable distance from the ganglion cell (vide fig. 35, *sf*). They have a dark staining; sometimes it looks as if they united to form a deeply staining peripheral layer round the contents of the process inside the sheath (fig. 36).

The sheaths of the nervous processes are formed by the neuroglia, and are stained in the same manner as this substance. They have generally a peculiar structure, and consist often of a great many layers or membranes aranged concentrically round the contents of the process; only the innermost layer of these membranes can, however, be considered as the real sheath. What a great extension these structures can obtain is, for instance, seen in fig. 34, where *pc* is the contents of a nervous process, *br* is a branchlet issuing from the process; *nm* the surrounding concentrically arranged neuroglia-membranes.

The nervous pocesses are sometimes provided with similar concentrically arranged sheaths quite from their origin in the ganglion cell (fig. 25, *sh*). Generally, the sheaths do not, however, reach their highest thickness before some distance from the cell, and they do not consist of so many layers near their origin.

The nuclei of the ganglion cells are, as is a well-known fact, enclosed by a thin membrane. That this membrane is not seen in the live-state of the cell, as stated by FREUD, is, in my opinion, from the same cause as makes the striation of the nervetubes so slightly visible in the live-state. The refractive difference between the protoplasm of the ganglion cell and the substance of which this membrane consists is so small, that this difference must necessarily be increased, for instance, by chemical agents, if the membrane should become distinctly visible (vide fig. 23). With hæmatoxylin this membrane gives a distinct blackish staining (fig.s 37, 38, 40, 42) which is quite similar to that of the spongioplasmic reticulation, and to the sheath of the cell. In my opinion, it is not improbable that this membrane is also formed by the same spongioplasm as forms the reticulation in the protoplasm and the sheaths of the primitive tubes.

The structure of the nuclei inside this membrane varies very much. As already stated by FREUD, there appears within the nucleus a great many changes, probably connected with the life of the cell and the nucleus. In well fixed preparations we can, in fact, also find nuclei exhibiting the greatest variation in their structure. Fig. 41 and 42 represent nuclei in various stages. Generally, each nucleus has one nucleolus situated in its centre or also near its membrane (cmfr. fig. 38). Sometimes two nuclei are present; the nucleus has, then very often, an oblong form, and one nucleolus is situated in each pole (fig. 37); it is, however, not unusual that the nucleus has a circular shape, and that nucleoli are situated oposite to each other, near the membrane (fig. 41). Sometimes no real nuclei are present, the whole nucleus exhibiting a peculiar distinct reticulation with dark thickenings in the points where the fibres of the reticulation unite (fig. 42, *d*). Sometime fibres are seen radiating from the centre of the nucleus, and having dark thickenings at their extremities, near the membrane of the nucleus (fig. 42, *f*) etc. etc. I will, however, pay no more attention to these structures, here, as they are probably only produced by the nuclei being fixed in different stages of their life; in some cases we have even, perhaps, a beginning segmentation of the nucleus (fig. 42, *e*).

Nephrops norvegicus.

In Nephrops, the structure of the ganglion cells is quite similar to what we have described of Homarus. In sections, the protoplasm of the large cells generally exhibits quite similar light areas, resembling vacuoli, and arising from the transsection of bundles or masses of primitive tubes. The contents of the nervous processes originate in the same manner; a similar spongioplasmic reticulation is present in the protoplasm, it being especially visible near the sheaths of the cells with which it is intimately united. We can, thus, as regards the structure of the ganglion cells of Nephrops, refer the reader to our description of the ganglion cells of Homarus.

The Polychætes.

Sections of the large ganglion cells of the *Nereidæ* (N. virens etc.) exibit, generally, a prominent granular appearance with a distinct reticulation extending between and uniting the granules with each other, the granules being generally situated at the points where the walls of the meshes unite (fig. 43). Some of the granules are very large and prominent, and are principally situated in the mesial part of the protoplasm. They have, in my preparations treated with fluids containing osmic acid and stained with hæmatoxylin, a very dark almost black staining, and they consist, I think, of a fattish (myeloid?) substance.[1]

In a great many cells, I observed a differentiation in the staining of the protoplasm, somewhat similar to what is described of the cells of Homarus. A disposition to a darker staining of the mesial part of the protoplasm is very often observed; besides this the larger granules are also, as just mentioned, situated especially here which contributes somewhat to a darker appearance (vide fig. 44

---

[1] Whether the substance of these granules is the same, or has any relation to the substance which gives the nervous system of many Polychætes the well-known reddish or yellowish staining, I cannot say, as I have not yet very closely examined the ganglion cells of the Polychætes in the live-state.

and 45). There is generally no distinct demarcation betwen the deeply stained mesial part of the protoplasm and the lighter peripheral layers; sometimes, however, such a demarcation is present, the lighter stained substance being peripherically situated in various more or less distinctly defined masses (fig. 44). Small bundles of primitive tubes, similar to those which are often present in the ganglion cells of Homarus and Nephrops, I have not generally observed in the ganglion cells of Nereis. The contents of the nervous process seems generally to originate in a gradual convergence of primitive tubes in the protoplasm, towards the pole where the process originates (cmfr. fig. 43). Sometimes, I believe to have observed the contents of the nervous process passing as an undivided, lightly stained bundle for some distance into the protoplasm; consequently, a structure somewhat similar to what is above described in the ganglion cells of Homarus.

A spongioplasmic reticulation quite similar to the peculiar reticulation in the ganglion cells of Homarus does not occur in Nereis, so far as our experience goes, at all events not to such an extent. In some cells, I have, however, observed spongioplasmic fibres issuing from the neuroglia-membrane enveloping the cell and penetrating into the protoplasm (vide fig. 43, *sf* and fig. 44, *sf*); probably contributing to form the reticulation already described.

A question of great interest is, whether the reticulation, seen in the sections of the cell-protoplasm of Nereis is a real reticulation or only an appearance produced by transsection of primitive tubes? I feel most disposed to believe that the latter is the case. Strange to say, neither longitudinally transsected primitive tubes, nor a concentric striation are as often seen in the ganglion cells as we could expect if all meshes were transsected primitive tubes; sometimes we can, however, observe longitudinally transsected tubes in the protoplasm, and the reason why they are not oftener seen is, perhaps, because they have a very curved and complicated course.

The nuclei of the ganglion cells have, in the Nereidæ, a similar structure as those of Homarus. They have a distinct surrounding membrane, inside which a reticulation is extended; in this reticulation dark granules are situated, generally at the points where the fibres forming the meshes unite. One or sometimes two nucleoli are seen, and sometimes no distinct nucleolus is present, but only largish granules are spread in the reticulation (vide fig. 43, 44, 45).

The situation of the nuclei in the protoplasm is generally different from what it is in the ganglion cells of Homarus. The nuclei are very often situated in one side of the cell, near the surrounding sheath (cf. fig. 44 & 45) and, not unusually, towards that side from which the nervous process issues (cf. fig. 43, 1). They are, thus, very frequently situated outside the mesial, deeply stained, part of the protoplasm (fig. 44 & 45). In the small cells they are generally situated in the centre of the cell, and are proportionally very large, being surrounded by a thin layer of protoplasm only.

The processes of the ganglion cells. — The ganglion cells of the Nereidæ have generally a unipolar shape. Quite exeptionally, I believe to have observed, in sections, cells with multipolar shape; there was always, however, only a single process in each cell capable of being traced in its course into the central dotted substance. The other processes were very short, immediately tapering off; they were directed peripherically, or laterally, (never towards the central mass) and were soon lost in the neuroglia. I could, however, never quite convince myself of the real existence of such processes, and it is only in a very few cases I believe to have seen them (fig. 67). I think that it is the issuing of neuroglia-fibres from the sheath of the cells that has generally occasioned the descriptions of multipolar cells, as such fibres can look very like real cell processes (cf. fig. 44).

The processes, or, as I call them, *nervous processes*, passing to the central dotted mass, and of which each ganglion cell has only one, have a structure similar to that of the nerve-tubes. They are surrounded by a neuroglia-sheath, and their contents consist of primitive tubes.

In other forms of Polychæta the ganglion cells have, so far as my investigations go, a structure similar, principally, to what is here described.

The membranes enveloping the ganglion cells are formed by the neuroglia. They generally consist of one, or of a few, layers only. In the small cells they are thin, but distinctly marked; in the large cells they are often thick and well developed (vide fig. 44, *cm*). Neuroglia-nuclei adhering to them do not occur very frequently.[1]

---

[1] Dr. *E. Rohde* (l. c. 1886 p. 785) distinguishes between two types of ganglion cells in the central nervous system of the *Aphroditidæ*. »Die Ganglienzellen der einen Art sind sehr schwach granulirt, deshalb von hellen Aussehen und meist ziemlich klein. Ihr Kern enthält stets mehrere verschieden grosse Körperchen und tritt nach Färbungen in der durchsichtigen Ganglienzelle scharf hervor.« »Die Vetreter des zweiten Typus sind sehr grosse, kugelige Gebilde, welche durch eine sehr dunkele Granulirung sofort in die Augen fallen. Sie besitzen einen grossen,

Lumbricus agricola.

The ganglion cells of Lumbricus are, in their structure, somewhat similar to those of Nereis. In sections, their protoplasm has a distinct, reticular, granulous appearance; large granules, similar to those found in the ganglion cells of Nereis, do not, however, occur. The whole protoplasm is, in preparations, rather deeply stained; the reticulation is generally very distinct, slight thickenings are only seen in the points where the walls of the meshes unite. The meshes are generally very large; that they are, at all events to a great extent, really transsected tubes is, I think, clearly seen in fig. 47, which represents a section through a large ganglion cell, and which is drawn under the camera lucida as exact to nature as possible (very highly magnified). The tubes are, in this illustration, seen transversally as well as partly longitudinally transsected, giving, the protoplasm to some extent, a concentric appearance; the tubes forming the nervous process are seen to be quite like those circulating in the protoplasm.[1])

In some cells, a kind of higher differentiation seems to be present in the protoplasm; such a cell is represented in fig. 46. Lighter

fein granulirten Kern und dieser ein einziges grosses Körperchen.« A similar difference in the appearance of the small and large ganglion cells is, certainly, very prominent also in the nervous system of *Nereis*; but as I have found transition forms between both these kinds of cells, I can not distinguish between two types. My experience is to some extent that, the larger the cells are, the more and more granular become their protoplasm, and there can not easily be drawn any line of demarcation.

As to the sheaths of the ganglion cells *Rohde* says: »Beide Ganglienzellenarten entbehren einer Zellenmembran und liegen eingebettet in ein Maschenwerk von Fasern, welche überall das Nervensystem begleiten und, wie ich glaube, aus Subcuticularzellen hervorgegangen sind.« As may be seen from my description I agree with this statement of *Rohde*; his »Fasern« or »Subcuticularfasern« belong to what I call the *neuroglia*.

Of the protoplasm of the largish cells *Rohde* says: »dass die ganze Zelle nach allen Richtungen von veschieden starken Fäserchen durchzogen wird, welche auf den Zellfortsatz übergehen und diesem eine feine Längsstreifung verleien. Aber nicht nur hier verlassen diese Fäserchen die Zelle, sondern man ist überrascht zu sehen, wie sie theils einzeln, theils zu Bündeln vereinigt allenthalben an der Peripherie des nackten Zellkörpers heraustreten und in die Subcuticularfaserhülle eindringen.« It is evident that *Rohde* and I have observed the same structures; his »Fäserchen« must partly be what in my opinion are the spongioplasmic sheaths of the primitive tubes (especially those which pass into the nervous process), and the »Fäserchen« penetrating in the Subcuticularfaserhülle« must be the spongioplasmic fibres issuing from the neuroglia-sheath of the cell (vide fig. 44, sf; fig. 45, sf; fig 67). and which fibres partly form a spongioplasmic reticulation in the protoplasm (compare my description of Homarus p. 101).

[1]) The tubes in the protoplasm seem to be of different size in the various cells.

and darker parts are here seen in the protoplasm, the latter ones forming a sort of reticulation with very large meshes, in which smaller meshes occur, similar to those mentioned above. This is perhaps an appearance produced by a structure somewhat similar to that which is described in the ganglion cells of Homarus (cf. fig. 25 & 27). The large meshes, or lightly stained areas, are perhaps transsected bundles or small masses of primitive tubes.

Such a prominent spongioplasmic reticulation as we have found in the ganglion cells of Homarus is not present in the ganglion cells of Lumbricus, indeed, it is very difficult to detect anything similar to it.

We have thus, the protoplasm of the ganglion cells of Lumbricus, according to my investigations, consisting principally of primitive tubes, with distinctly marked spongioplasmic sheaths; besides this there must, however, also be another subtance present, as the protoplasm is more deeply stained then the contents of the nerve-tubes. This protoplasm has therefore, in Lumbricus, a composition which probably is very similar to what is present in the ganglion cells of Homarus.

The processes of the ganglion cells. — The ganglion cells are unipolar or multipolar. Each cell has, however, never more than one *nervous process* passing to the central dotted mass of the nerve-cord. The other processes are very short, have the same appearance as the protoplasm of the cell, and are lost in the surrounding reticular neuroglia, or can be traced to their connection with the neurilem-sheath or *perineurium* surrounding the ventral nerve-cord inside the muscular layer (vide fig. 68). These processes I call *protoplasmic processes*, and their function is, I suppose, of nutritive nature. All ganglion cells are thus, in reality, of unipolar nature.

The structure of the nervous processes is similar to that of the nerve-tubes; their contents consist of primitive tubes, and they are eveloped by a neuroglia-sheath which is a direct continuation of the membranes or sheaths enclosing the ganglion cells from which they issue.

The nuclei of the ganglion cells are comparatively large. They have the same appearance as described above, with a distinct surrounding membrane and a varying inner structure. Their situation in the cells is also somewhat variable; they are very often situated towards that pole from which the nervous process issues; this is perhaps the most common situation (fig. 46); sometimes they are, however, also situated towards the opposite end of the cell, or in its mesial part (vide fig. 47).

The membranes enveloping the ganglion cells are formed by the neuroglia; they are generally very thin and only little prominent. Occasionally, neuroglia-nuclei are seen adhering to them. They are intimately connected with the reticulation of neuroglia extending between the ganglion cells (vide fig. 47 & 68).

The Mollusca.

The structure of the ganglion cells af *Patella vulgata* is very similar to what we have found in *Lumbricus*. Their protoplasm exhibits, in sections, a distinct reticulation with rather largish meshes (vide fig. 48, 49 and 50). In this reticulation small granules are seen, situated in the same way as described above, viz. in the points where the walls of the meshes unite. The meshes are, to a great extent, transsected tubes, this is seen in several cells where some tubes are longitudinally transsected.[1])

In macerated preparations plenty of large yellow granules are generally seen in the protoplasm of the ganglion cells. These granules have a variable size, and no regular shape, they are sometimes spherical, sometimes square or polyhedrical, and they look as if they were produced by coagulation of a homogeneous yellow substance. They are often extended through the whole mass of the protoplasm, very frequently they are, however, concentrated in special parts of the cells, especially in the neighbourhood of the nucleus. Plenty of similar smaller or larger granules generally occur, also, outside the ganglion cells, in macerated preparations. They frequently occur in such number that one, for a time, could feel disposed to believe that they belonged to a substance extended through the whole nervous system. Sometimes they are even united to larger homogeneous masses. Upon careful examination I have, however, come to the conviction that they are, either exuded from cells, or they may also spring from destroyed cells. I have sometimes observed such a substance exuded from the protoplasm of cells.

Fig. 52 represents such a case. The substance is here seen occurring inside, as well as outside, the cell. Inside the cell, the granules are concentrated towards the part of the surface where they

---

[1]) The reticulation which *Rawitz* describes, in the ganglion cells of the *Acephales* (vide p. 65) is, without doubt, the same appearance as here described, and it is, consequently, in my opinion, no real reticulation but, to a great extent, an appearance produced by primitive tubes, which are seen in optic, or real transverse, sections.

are probably exuded, outside the cell they are united to larger pieces of irregular shape. The granules are not only situated nearthe surface of the cell, but also occur in the mesial parts of the protoplasm; this can easily be seen by changing the level of the microscope.

The relations of the substance of these yellow granules in the live-state of the cells, I have not sufficiently examined, neither am I in a position to say of what it consists. It is obviously this substance which gives the nervous system of *Patella*, as well as other Molluscs, the well-known yellowish colour, and which RAWITZ, HALLER, H. SCHULTZE, SOLBRIG, BUCHHOLZ and others have described as pigment. I think it is most probable that the yellow colour is due to a substance, related to, or similar to hæmoglobin; the substance contains, probably, also fat, as the granules are deeply stained by osmic acid. Strange to say, I have, as a rule, not been able to observe quite similar granules in the sections of the cells. The reason why I do not exactly know, as my investigations are still too imperfect in this respect.[1]) As also suggested by HALLER, I think it is very probable that this substance is principally engaged in the nutrition of the ganglion cells. I suppose that the corresponding substance in the ganglion cells of Homarus is the fatty substance, the existence of which we have above (vide p. 108) indicated.

In the sections of some cells, I have observed a slight tendency to a differentiation of the protoplasm into darker and lighter stained parts (vide fig. 48). This tendency is, however, never very prominent. Generally, the protoplasm has a rather uniform, somewhat deep staining, which possibly indicates the extension of a (myeloid?) substance, connected with the spongioplasm, through the whole protoplasm of the ganglion cells, as before mentioned in my description of the ganglion cells of Homarus, and possibly this substance is of the same nature as that which forms the granules (above mentioned) in macerated preparations.

The processes of the ganglion cells. — The ganglion cells have a unipolar, bipolar or multipolar shape. Whether unipolar or multipolar, each cell has one *nervous process* only, passing into or through the central dotted substance, the other, *protoplasmic processes*, if they are present, are always very short, and have an appearance similar to the protoplasm of the cell. They are immediately lost in the neuroglia-reticulation, or they pass to, and unite with the perineurium surrounding the central nervous system.

---

[1]) In a great many sections I have found a homogeneous yellow substance being extended in the protoplasm of the ganglion cells. It seems to be the same substance which forms the granules in isolated cells.

system. Their function is, in my opinion, of nutritive nature, as mentioned before.¹)

The contents of the nervous processes consists of primitive tubes.

The nervous processes (as well as the protoplasmic ones) are enveloped by neuroglia-sheaths adherent to which neuroglia-nuclei may be seen.

The nuclei of the ganglion cells have a structure similar to what is before mentioned as regards the nuclei of the cells of Homarus etc.

They are generally situated in, or towards, that pole of the cell from which the nervous process issues. Sometimes they are also situated in the mesial part of the cell and very seldom in the end opposite to the nervous process.

They have a relatively large size; when they are situated near the origin of the nervous process, they generally fill nearly the whole diameter of the cell, leaving only a narrow layer of protoplasm, through which the connection of the nervous process with the upper principal part of the protoplasm is produced (vide fig. 53). Frequently the space between the nucleus and the membrane enveloping the cell is, however, so extremely narrow, that it looks as if only a very few primitive tubes could pass; it has, indeed, the appearance as if the protoplasm of the ganglion cell was almost divided by the nucleus into two separate parts.

In one case I have seen a nucleus sending a short process into the nervous process (vide fig. 54); this short process was very distinct, and appeared to be formed by the nuclear membrane; it may be that it has only been an artificially produced, post-mortem, appearance.

The neuroglia-membranes enveloping the ganglion cells are thin and but slightly prominent. Neuroglia-nuclei, adherent to them, are occasionally seen. They are intimately connected with the neuroglia reticulation extending between the ganglion cells (fig. 48—50, *nur*). When HALLER and RAWITZ seem to deny to some extent the existence of such membranes, I think that it is owing

---

¹) In the peripheral layers, almost all cells have unipolar shape, in the inner layers (towards and in the dotted substance), the bipolar or multipolar shape is more common; as will later, in a separate chapter, be treated of, this also indicates the nutritive significance of the protoplasmic processes. Real anastomoses of protoplasmic processes from the same cell, or from various cells, I have never been able to observe, and do not believe either in their existence or in the significance which *Haller* and most writers attribute to those processes (vide sequel).

to imperfect staining. As above mentioned, these membranes also envelope the processes issuing from the cells.

The Ascidians.

The protoplasm of the ganglion cells of the Ascidians has, in isolated, macerated preparations, a reticular appearance (vide fig 55—56), quite similar to what is before described in *Mollusca*, *Lumbricus* etc., and which appearance is, I suppose, to a certain extent, produced by primitive tubes, which in a complicated way are woven between each other and are, in the preparations, partly seen in optic diameter. It is remarkable that a concentric arrangement in the protoplasm round the nucleus can, only very seldom, be traced out. The shape of a great many cells, and the situation of the nuclei is, perhaps, to some extent, the cause of this. In sections, the protoplasm of the cells exhibits a reticulation similar to what is observed in isolated cells. At the points where the walls of the small meshes unite, small thickenings or granules occur, giving the protoplasm a slightly granular appearance.

I have never observed in or outside the ganglion cells of the *Ascidians*, yellow granules, similar to those just described in the ganglion cells of *Patella*, neither in macerated preparations nor in sections.

In sections of some large cells, I have observed a tendency to differentiation into small lighter stained areas, similar to what is described in the ganglion cells of Homarus. This differentiation is not, however, very prominent (vide fig. 58, *a*).

In preparations treated with osmic acid, or fluids containing this agent, the whole protoplasm of the ganglion cells is also very deeply stained by carmine colours (picro-carmine).[1]) This indicates perhaps the presence of a special substance (compare what is said of this above.)

The nuclei of the ganglion cells have a distinct thin membrane, and an inner structure which has a varying appearance, similar to what is above mentioned in respect of the ganglion cells of Homarus etc. A distinct nucleolus is generally seen. The nuclei are relatively large; they are situated in the mesial parts of the cells,

---

[1]) A method which in my experience is very good for the nervous system of the Ascidians, is treating with osmic acid ($\frac{1}{2}$—1 $^0/_0$) for $\frac{1}{4}$—1 hour, or even longer, then sufficient washing in running water and staining in good picro-carmine (1 $^0/_0$) for 24 hours or longer.

or also sometimes near the pole from which the nervous process issues or in the opposite side.

The processes of the ganglion cells. — The most common shape of the ganglion cells is the unipolar one; the large, peripherically situated cells, especially, are of this shape (fig.s 55, 56, 58). In the inner layers and in the dotted substance, cells with bipolar, tripolar or multipolar shape occur (fig. 57). Each cell has only one *nervous process*, the other processes are *protoplasmic*, and their function is of nutritive nature. That is also, in my opinion, the reason why they principally occur in connection with cells situated in the inner layers, and are directed towards the external sheath (*perineurium*) enveloping the brain.

The *nervous* processes exhibit a longitudinal striation, and their contents consist of primitive tubes. They are, in macerated preprations, only lightly stained, and have an appearance different from that of the protoplasm of the cells. The protoplasmic processes resemble, in their appearance, the protoplasm of the cells.

The neuroglia-membranes enveloping the ganglion cells are very thin and but slightly prominent. The neuroglia is but little developed, and occurs very sparingly between the ganglion cells in the brain of the *Ascidians*.

Summary.

If the reader has followed me in these researches on the structure of the ganglion cells of various types of invertebrates, he will have gathered that there are some principal features in the structure which seem to be common to all the types investigated; and, if it is so, we may conclude as above, in respect of the structure of the nerve-tubes, that those features in the structure of the ganglion cells are common to all invertebrated bilaterates. We may thus give the following summary of our results:

1) The ganglion cells of all *invertebrated bilaterates* consist of a *nucleus* with distinct membrane, and a varying inner structure, and also of a *protoplasm* with various constituents; the cells are enclosed by a membrane of neuroglia-substance.

2) The principal constituents of the protoplasm are *primitive tubes*, having the same structure as those in the nerve-tubes (they contain *hyaloplasm* enclosed in *spongioplasm*). The course, and origin (or termination) in the cell, of these primitive tubes I am not in a position to describe particularly; some of them very frequently circulate

for some distance, concentrically, round the nucleus, giving the ganglion cells a concentrically striated appearance.

In some ganglion cells, especially those of *Homarus* and *Nephrops*, primitive tubes are partly united in bundles, or to smaller or larger masses situated in the protoplasm, and which are distinctly lighter staining than the rest of the protoplasm, in which, however, also, plenty of primitive tubes occur (fig. 38).

In a great many, or possibly in all, ganglion cells a *spongioplasmic reticulation* is present, extending from the enclosing neuroglia-membrane into the protoplasm, between the primitive tubes, and intimately connected with the spongioplasmic sheaths of the latter. This reticulation is very prominent, especially in the ganglion cells, of Homarus and Nephrops, where thick spongioplasmic fibres af a peculiar appearance, and connected with the reticulation often occur in the peripheral layers of the protoplasm, penetrating from the neuroglia-membrane into the protoplasm (fig. 24).

Besides this reticulation, there is, probably, also a special, partly fatty (myeloid?) substance present in all ganglion cells of invertebrated bilaterates, which substance does not generally occur in the nerve-tubes. This substance possibly occurs to some extent, in connection with the spongioplasmic reticulation, and extends between the primitive tubes of the protoplasm, giving the latter, in preparations, the deep staining which is generally prominently different from the staining of the contents of the nervous processes and nerve-tubes where primitive tubes only occur.

Whether it is the same substance which, in the ganglion cells of a great many animals (cf. Polychæta, Mollusca etc.), is connected with a pigment (hæmoglobin?) and gives the nervous system its special colour, I am not in a position to decide, although I think it not improbable.

3) The processes of the ganglion cells are of two kinds; viz. *nervous processes* and *protoplasmic processes*.

Of *nervous processes* each ganglion cell in the central nervous system has *always one* and *never more*. The nervous processes are generally directed, centrally, towards the dotted substance.

When the ganglion cells are multipolar (the unipolar cell is, however, the most common type in invertebrated bilaterates) the other processes are *protoplasmic ones*, and they are generally short and directed peripherically, they having a nutritive function and being united with the neuroglia. As regards structure and appearance, they are quite similar to the protoplasm of ganglion cells.

4) The contents of the nervous processes consists of primitive tubes which spring from the protoplasm of the ganglion cells, generally in such manner that they converge uniformly from the whole protoplasm, towards the pole where the nervous process issues; here they unite, and constitute the contents of the latter (fig.s 26, 43—58).

In some ganglion cells (observed in Homarus and Nephrops) the contents arises from a union of bundles of primitive tubes. The contents of the nervous process may also be formed, in this or the common manner, within the protoplasm of the cell for a shorter or longer distance from the place where the process issues, the process-contents has thus, to some extent, an undivided course through the protoplasm within the cell itself (fig.s 37, 38 A, 59, 60). These latter modes of origin I have observed especially in Homarus and Nephrops.

3. The structure of Leydig's dotted substance.

The nature of the so-called LEYDIG's dotted substance (»Punktsubstanz«), which is centrally situated in the nervous ganglia or the central nerve-system of all invertebrated bilaterates, has been very much discussed, and has been described in very different ways by the many previous authors on this subject, as is already mentioned in the historical introduction to this paper.

Where so many prominent scientists have given their opinion, it is, of course, a serious matter to suggest a new view which is contrary to almost all previous views.[1] It was therefore not without some hesitation and only after careful investigations that I entered upon the description of the dotted substance in the nerve-system of the *Myzostomes*[2] and subsequently upon the dotted substance in the brain of the *Ascidians*.[3] Since that time I have extended my investigations to a great many animals of various classes, and have always found my previous results confirmed in the principal respects.

My investigations lead me, thus, to maintain the view I have previously stated, and I am now in a position to state it, as I

---

[1] *Leydig* is, as before mentioned, the author with whom I can agree in most respects.
[2] l. c. 1886.
[3] l. c. 1886.

nope, with better and more complete expression. As will be seen from the following description, the structure of the dotted substance is essentially the same in all the animals investigated. The animal in which I have found it easiest to get a clear idea of the structure is Homarus, the ventral ganglia of which give an excellent material for investigations of this nature.

Homarus vulgaris.

Sections through the brain or the ventral ganglia of *Homarus* exhibit, in their mesial part, a more or less minute reticulation; to some extent this reticulation even passes over into masses which, on a superficial examination, have a dotted appearance, for which reason LEYDIG has also called it »the dotted substance« (»Punktsubstanz«), without intending, however, that this designation should be understood in its literal signification. On a more careful examination of good and successfully stained sections, even these masses with the finest granular appearance exhibit a reticulation with very minute but still distinct meshes; the granules are thickenings in the reticulation. On examination of sections through a ventral ganglion we will, indeed, find reticulations with meshes of all possible sizes, from the transsected large nerve-tubes (dorsally situated and issuing from the longitudinal commissures or from the peripheral nerves) down to the extremely minute meshes in the dotted masses just mentioned, which meshes are of about the same size as the transsected primitive tubes in the sections of the nerve-tubes or even much smaller.

The substance forming this reticulation is rather uniformly stained through the whole mass, and in a manner similar to that in which the neuroglia-sheaths of the nerve-tubes are stained (vide fig. 62). It is a distinct staining which is, consequently, different from that of the spongioplasm in the nerve-tubes (vide fig. 62). This makes me believe that it is not quite the same substance as that which I have, previously, in this paper called spongioplasm, but that it is rather the neuroglia-substance which forms this reticulation, as I can, really, see no distinct difference between the neuroglia enclosing the smaller or larger nerve-tubes and the substance forming the smallest meshes.

Within the meshes a lightly stained, hyaline, substance occurs, which is similar to the hyaloplasm of the nerve-tubes.

A question of great interest is now, whether these small meshes, seen in sections, belong to a *a real reticulation* formed by *fibrillæ*, as almost all writers agree in describing it (some writers call it

»nervous retic.« others call it partly »retic. of connective-tissue«) or whether they are transsected tubes similar to the meshes produced by the transsection of the larger nerve-tubes with which we are already acqvainted. In the former case, the hyaline substance seen within the meshes should be interfibrillar substance, in the latter case it must be a substance filling the tubes, probably hyaloplasm.

On careful examination of transverse sections we will immediately receive a reply to this question. If the meshes are transsected tubes, we may expect to find in a section through a mass consisting of a plait of such slender tubes, not only transversally transsected tubes but also longitudinal transsected ones. And that is, in fact, the case. On a glance at fig. 62 (which represents a part of a transverse section through a ventral ganglion of *Nephrops norvegicus*, which is, however, so quite similar to Homarus, in this respect, that we can indeed see no difference) we will, in the fine reticulation (*ds*) be able to see transversally transsected tubes as well as longitudinally transsected ones, the latter having the shape of more or less oblong meshes.

On examination of spots where small parts of nerves originate in similar masses of dotted substance, its composition of tubes will be still more evident. Fig. 61 represents such a spot, highly magnified, in the mesial part of the first ventral ganglion of Homarus. That the meshes, *tpt*, are transsected tubes is, I think, distinctly seen; *c* is a tube partly longitudinally transsected; *a* is a bundle of similar tubes issuing from various parts of this mass of dotted substance and passing to the root of a nerve. Many similar proofs of the tube-nature of the meshes, seen in the dotted substance, can be found on examination of sections through the ganglia of Homarus, and each of them speaks so clearly that I think it, really, to be a waste of time to give further, circumstantial, description of it at present.[1])

Having thus elucidated this question regarding the tube structure of the dotted substance, the next question of interest becomes — of what kind these tubes are, whether nerve-tubes or primitive tubes, or what else?

Nerve-tubes are, as previously mentioned, present in great plenty in the dotted substance, as will be seen in fig. 62—65. These nerve-tubes have all possible gradations from large, fig. 62, *t nt*' and *t nt*, down to very small ones, of which we have, for instance, a transsected bundle in fig. 62, *s nt*. The smallest meshes or tubes, *ds*,

---

[1]) In some parts of the dotted substance the tubes are somewhat loosely situated, and the intervals between them are then filled with a *neuroglia sponge-work*.

are, however, all of them, smaller than those small nerve-tubes, and they have, to a certain extent, a rather uniform size, with their diameter about the same as that of the common primitive tubes of the nerve-tubes (vide in *t nt'*). There are, also, a great many extremely minute meshes or tubes which are even smaller than the smallest primitive tubes I have been able to observe in the nerve-tubes. These minute tubes will, subsequently, be more circumstantially described. The principal difference between the primitive tubes of the nerve-tubes and the tubes of the dotted substance is that, the latter have more distinct and deeper stained sheaths. Seeing how nerves are formed by the union of these tubes, it is, therefore, in my opinion, evident that they are primitive tubes with stouter sheaths than they have in the nerve-tubes. These, the sheaths, in the dotted substance, seem to be formed or, at all events, made stronger by the same neuroglia which also envelopes the nerve-tubes; this neuroglia does not seem, however, to be very distinctly distinguished from the spongioplasm, as we have, also, previously seen in the ganglion cells (cf. p. 100—102). Inside these primitive tubes, I have been able to observe any structure, only the hyaline substance, hyaloplasm, above mentioned. The granules which have given the dotted substance its name, are, partly, thickenings in the sheaths of the tubes, especially along their concreting edges; to some extent they are extremely slender tubes traussected (vide sequel).

The meshes which are somewhat larger than those small meshes in the dotted substance, are, I think, transsected small nerve-tubes consisting of a few primitive tubes only.

We have thus, in the dotted substance in the ganglia of Homarus, a complicated plaiting or web of primitive tubes, and partly of nerve-tubes, of various sizes. On examination of longitudinal sections we will find that, these tubes have in the dotted substance of the ventral ganglia, to a certain extent, a tendency to follow a longitudinal course; this is especially the case in the parts where the nerve-tubes issuing from the longitudinal commissures are situated. To some extent, we also find bundles of nerve-tubes running transversally as commissures from one side of the central nerve-system to the other, or also running to the peripheral nerves.

Smaller or larger distinctly defined masses, apparently consisting exclusively of primitive tubes, are mesially situated in most ganglia; they are, however, especially prominent in the first ventral ganglion and in the brain; fig. 61 represents, as before mentioned, a part of such a mass situated, mesially, in the first ventral (thoracic) ganglion.

The illustration is drawn under the camera lucida and very high microscopical power.

As we can not undertake the very complicated topography of the ganglia of Homarus in this paper, we will not enter upon the topographical peculiarities here, which KRIEGER, YUNG, DIETL and others have already to some extent previously described; we must confine ourselves to indicate the constituents of the various masses of the »fibrillar« substance (dotted substance) filling the mesial parts of the ganglia. The constituents of these we have found to be *primitive tubes*, or also *nerve-tubes*, and *neuroglia*.

Though I have not, as mentioned, been able to observe any structure inside the primitive tubes described there may perhaps be a still smaller or more minute constituent in the dotted substance. Such minuter constituent I have, however, not observed in common preparations stained in hæmatoxylin, carmine etc.; it is only on application of the chromo-silver method, mentioned p. 77—80, and partly on staining by HEIDENHAIN's hæmatoxlin method, that it has been possible to observe such a constituent, and even then only imperfectly. It consists of extremely slender fibrillæ or rather tubes, which run in all directions in the dotted substance between the larger tubes, and whose diameters are much smaller than any of the primitive tubes described. Such fibrillæ or tubes are seen in fig.s 63—65. Many of them are, here, seen to be given off from larger nerve-tubes. Some of them have, at certain intervals, varioceles (vide fig. 63, *f*; fig. 64, *i*, *vf.*), and resemble in their appearance the varicose nerve-fibrillæ I have found in the central nerve-system of Myxine (and which will be subsequently described) and the varicose nerve-fibrillæ described by GOLGI in the central nerve-system of the Mammalians, and, further, the nerve-fibrillæ described by BELLONCI in the tectum opticum etc. of fishes and birds, etc. etc. In the varioceles extremely slender branches are probably given off, these have, however, only in a few places been stained. The varioceles exepted, the fibrillæ are smooth and have a deep reddish-black staining. As will be seen from the illustrations, their thickness is very variable; they subdivide and at each subdivision they grow thinner.

Whether the structure of those fibrillæ is that of tubes with sheaths and semi-fluid contents, as we have previously described the primitive tubes to have, is of course extremely difficult to decide. We know, at present, so very little of the nature of the chromo-silver staining, that it can only give us little instruction in this respect. The reason why these fibrillæ, and also small nerve-tubes, are so distinctly and deeply stained, whilst the sheaths of the larger nerve-tubes only get a light reddish staining is, for instance, very difficult to explain. Seing, however, that it is only the sheaths of the larger nerve-tubes or a layer just inside the sheaths which becomes stained, we may perhaps conclude that this is also the case with the smallest nerve-tubes; indeed, we can, to some extent, observe it in transverse sections of them, though their contents also appear to be somewhat stained. Seeing that it is the case with the smallest nerve-tubes from which these fibrillæ issue, and which they quite resemble in their staining, it is, in my opinion, most probable that they also have an external layer which is the essential staining part of them, notwithstanding that the contents are also stained. That it is principally the external layers of the fibrillæ which are stained, can sometimes be observed, especially in the thicker fibrillæ, or in their varicose thickenings. To speak more distinctly I will say that I do not think it is the external part of the neuroglia-sheaths, but that it is either their internal parts or rather a layer inside them which is specially staining in these slender tubes.

Judging from chromo-silver preparations I think therefore that it is probable these fibrillæ are tubes, the smallest of which must consequently have an almost infinitesimal diameter. On examination of preparations stained with HEIDENHAIN's hæmatoxylin method we arrive at very similar results; indeed, the tube-structure of these slender fibrillæ (which are also partly stained by this method) is still more evident. Fig. 65 represents a part of a section through such a preparation of the dotted substance (drawn under the cam. luc. and very high powers of the microscope). The slender fibrillæ, which are deeply stained, are seen longitudinally or transversally transsected in this very thin section. Transversally transsected they appear, partly, as extremely minute meshes, which we have before mentioned. Longitudinally transsected they, partly, exhibit very distinct longitudinal outlines and a lighter contents. Frequently they are, however, so slender, that no such structure can be distinguished, neither in transverse sections nor in longitudinal ones. Of the tube-structure of these smallest fibrillæ I must, therefore, once more expressly say that we do not yet know anything with certainty.

Another question is, whether are those fibrillæ or tubes only very thin primitive tubes which by subdivision etc. have become so thin, or are they a special constituent contained in the primitive tubes? Seeing that they issue from nerve-tubes, I think it is evident that the former, or the latter must be the case, *tertium non datum*. That they are of real nervous nature, and are not simply fibres belonging to the connective substance, neuroglia, is not, I think, open to argument.

To decide the above question is not easy. As before mentioned, it has not been possible to detect any structure inside the sheaths of the primitive tubes, neither in the nerve-tubes nor in the dotted substance. In spite of this a structure may of course be present; the primitive tubes are already so very minute that even our present powers of the microscope, though high, would not readily suffice to exhibit such a structure of small tubes or fibrillæ inside them. We, therefore, at present, stand, here, before a terra incognita, and must content ourselves with suppositions, which we will, however, leave the reader to form for himself. What we know is that, these fibrillæ spring from subdivisions of nerve-tubes or primitive tubes — or they are given off, from them, in form of slender lateral branchlets, and it is then, perhaps, most reasonable to assume that they arise only by a subdivision of primitive tubes.

As to their course in the dotted substance, I will expressly say that, I have *never* succeeded in observing these *fibrillæ to form a reticulation with real meshes*, neither have I seen them *anastomose with each other*. They frequently exhibit, in sections, an extremely complicated course with a great many subdivisions and branches, but in my preparations they always avoid union with each other. They form, consequently, a kind of loose plaiting or web and not a reticulation as most authors describe. They pass along the walls between the thicker tubes of the dotted substance.

What previous authors have described as nervous reticulation in Homarus, as well as other invertebrates, is, as mentioned above, the transsected tubes, primitive tubes and nerve-tubes, forming the dotted substance, the sheaths of which tubes, in sections, give the appearance of a reticulation. LEYDIG describes, as mentioned p. 60, a sponge-work (»Schwammwerk«, »Balkenwerk«) in the dotted substance, which sponge-work he supposes to be of the nature of a support; the real nervous substance, hyaloplasm, is diffusively extended in the cavities of this sponge-work.[1]) As may be seen from

---

[1]) *Leydig* does not exactly state what he supposes to be the origin of this reticulation. In »Zelle und Gewebe« 1885 p. 173—174 he only speaks of the dotted substance as containing a »protoplasmatisches Netz« oder richtiger Schwammwerk«. Of this »Schwammwerk« or »Balkenwerk« he says: »wo nun Nervenursprunge gesetzt sind ordnet sich das Balkenwerk zu Längsstreifen, die zwischen sich die homogene Grundsubstanz ebenso aufnehmen, als es in dem sich durchkreuzenden Maschenwerk geschehen war.« In another place he says, however, that the dotted substance »entsteht durch fortgesetzte Theilung und netzige Auflösung der Fortzätze der Ganglienkugeln, genauer gesagt, ihres Spongiplasma« (l. c. p. 187).

my description, above, I agree principally with LEYDIG, as to the nature of the two substances, but we do not agree as to their structure. LEYDIG calls the substance of his reticulation *spongioplasm*; as I have tried to distinguish between *spongioplasm* and *neuroglia*, I have called the same substance *neuroglia*, which, however, in my opinion, forms tubes (enveloping primitive tubes or nerve-tubes) and not a sponge-work in the dotted substance. LEYDIG does not draw any line of demarcation between spongioplasm and neuroglia. His opinion is that, what he calls spongioplasm is a reticulated substance which is present in the ganglion cells, as well as in the cells of the neuroglia, or the connective-tissue as he calls it; and that there is, in the nerve-system of the vertebrates, an intimate connection between the spongioplasm of both kinds of cells (cf. op. cit. p. 187—189). In a future paper on the structure of the neuroglia, the writer will have an opportunity to treat of this subject more circumstantially.

The origin of the primitive tubes and fibrillæ of the dotted substance. — Having described what the constituents of the dotted substance are, as far as our ability goes, we will now advance to examine from whence these constituents come. To do this, we must try to learn the course of the nervous processes, issuing from the ganglion cells, and the nerve-tubes in the dotted substance.

We have, already, said that the nervous processes of the ganglion cells occasionally subdivide, and give off branches, on their course from the ganglion cells to the dotted substance. On a closer examination we will find that they do the same, in a higher degree, on their course through the dotted substance.

This subdivision and branching of the nervous processes cannot easily be traced, without staining by the *chromo-silver method* (cf. p. 78). In successful preparations, stained in this way, I have occasionally been able to trace the nervous processes, to some extent, on their course through the dotted substance. When a nervous process was visible for some distance along its course dichotomous subdivisions, or finer side-branches given off from it, were always observed; I have never observed a nervous process which, for any considerable length, had an isolated course through the dotted substance.

On comparison of the course of the various nervous processes, I have found that they essentially differ, and that there must be two kinds or types of them, which behave in two different ways on their course through the dotted substance.

In the course of some processes dichotomous subdivisions are very common, and the branches of the process subdivide, again, into smaller and smaller branches, this seems to continue until the whole process is broken up into a great many fine primitive tubes or fibrillæ, and *its individuality is*, consequently, *quite lost*. I have not, yet, been able to trace any process to its division into the finest tubes, but from the little I have seen, however, I believe that I am entitled to conclude that such must be the case. Fig. 63, *a* and *b* represent pieces of such processes, which are drawn under the camera lucida direct upon the stone; fig. 70 represents a ganglion cell with such a process.

A great many processes have quite another character. I have been able to trace them for long distances through the ganglia, in one case even directly into the root of a nerve, without seing any subdivision. They have, however, no isolated course; at certain intervals they give off slender side-branchlets which often subdivide in the dotted substance. At the places where such branchlets issue, the nervous processes have generally small thickenings or varioceles. I think that all the nervous processes of this type pass to a commissure or peripheral nerve, and become a commissural or peripheral nerve-tube. We may thus say, of these nervous processes, that they *keep their individuality, but have no isolated course*. Fig. 64, *a, b, c, d* represent nervous processes of this kind; fig. 68 and 69 represent ganglion cells with such processes.

We may thus establish two types of nervous processes viz. 1) *nervous processes* which lose their individuality and are entirely *broken up into slender primitive tubes* and *fibrillæ*, and 2) *nervous processes which keep their individuality* and pass through the dotted substance of the ganglia, *forming a nerve-tube*, but which *have no isolated course*, side-branchlets being given off on the way through the dotted substance.

We have, before, said that the nervous processes subdivide, and give off branchlets also before the reach they dotted substance. The branches and branchlets which arise in this way seem, however, chiefly, if not wholly, to penetrate into the dotted substance; they frequently enter into this substance together with the thicker nervous processes, as will be seen in fig. 64, *e, f*, where several such branches are represented. It seems, thus, not to be of any essential import whether the nervous processes subdivide in or outside the dotted substance, as in both cases the branches penetrate into it and become one of its constituents.

It is, however, not only the nervous processes which subdivide or give off branches to the dotted substance and thus contributes to its formation; the nerve-tubes coming from the longitudinal commissures and from the peripheral nerves also do the same.

That the *largish nerve-tubes of the longitudinal commissures* subdivide to some extent in the ganglia, may already be concluded from the fact that, in a transverse section through the central part of a ventral ganglion, only a very small number of transsected large nerve-tubes is seen. The many largish nerve-tubes of the commissures must, therefore, either have passed to the peripheral nerves, or they must have subdivided, or have become diminished by giving off lateral branchlets, or finally they may originate in ganglion cells. On examination of longitudinal sections stained with hæmatoxylin etc. it may be seen that these longitudinally running nerve-tubes subdivide, in the ganglia, to a great extent. The contents of the longitudinal commissures radiate into the dotted substance of the ganglia, and are to some extent lost in it owing to the subdivisions of the nerve-tubes.

On examination of longitudinal sections of preparations successfully stained by the chromo-silver method this is of course seen much more distinctly. In such sections, I have seen longitudinal nerve-tubes which were broken up by subdivisions into fine primitive tubes and fibrillæ (vide fig. 64, $_{1, 3, 11}$).

I have, however, also seen longitudinal nerve-tubes passing undivided through the ventral ganglia and into the commissures quitting them at the other side. Sometimes I have observed such nerve-tubes to give off side-branchlets to the dotted substance of the ganglia through which they pass (cf. fig. 64, $_{2, 6, 8, 9, 10}$).

I believe that some of the longitudinal nerve-tubes pass to ganglion cells, and are connected witht their nervous processes, or, in other words, that they are direct continuations of the nervous processes. I have certainly in no preparation succeeded in really observing such a direct connection; I have, however, seen so many indications of its probable existence that I do not think there can be much doubt about it. These nerve-tubes do not, however, have any isolated course, they give off side-branchlets to the dotted substance.

The nerve-tubes of the peripheral nerves originate in the ganglia in two, or rather three, ways.

Some nerve-tubes spring directly from ganglion cells, being direct continuations of nervous processes, as before mentioned. These nerve-tubes, or nervous, processes give off side-branchlets on their way through the dotted substance, vide fig. 69, which represents such a nerve-tube, seen in a section stained by the chromo-silver method; in the section a few lateral branchlets only were stained, but I think it probable that there were more. The nerve-tubes spring from ganglion cells which are situated either on the same side as the peripheral nerve to which the nerve-tubes belong, or on the opposite side of the ganglion.

Other nerve-tubes do not spring directly from ganglion cells, but *arise from the dotted substance by a union of slender primitive tubes and fibrillæ*, which unite to form thicker and thicker tubes. I have never succeeded in tracing a nerve-tube with such an origin to its finest branches; this is owing to the very complicated course of those nerve-tubes; they often originate on the same side as the peripheral nerve to which they belong, but most frequently they originate on the opposite side. They then pass united to bundles, or transverse commissures, from one side of the ganglion to the other. Though I have not been able to trace such nerve-tubes through the whole extent of their course, I have, however, seen their origin in portions and believe I am entitled to say that it is as just described.

When we now gather what will be the constituents of the dotted substance according to the above given investigations we will have the following.

1) The branches of those nervous processes which lose their individuality and are entirely broken up into fine branches.

2) The side-branchlets of those nervous processes which directly become nerve-tubes and do not lose their individuality (but which on their course through the dotted substance give off side-branchlets).

3) The branches of those nerve-tubes which come from the longitudinal commissures, and which in the dotted substance are entirely broken up into slender branches.

4) The side-branchlets given off from those nerve-tubes of the longitudinal commissures which pass entirely through the ventral ganglia, only giving off branchlets to the dotted substance.

5) The side-branchlets joining those nerve-tubes of the longitudinal commissures which spring directly from ganglion cells. These branchlets are partly the same as those mentioned in 2.

6) The primitive tubes and branchlets, or fibrillæ, which unite

to constitute those peripheral nerve-tubes which entirely spring from the dotted substance.

7) The side-branchlets joining those peripheral nerve-tubes which are direct continuations of nervous processes from ganglion cells. These side-branchlets are, consequently, partly the same as those mentioned in 2.

We have thus seen that the constituents of the dotted substance are tubes, and that these tubes have a rather variable origin. What the significance of the dotted substance is, we will in subsequent chapters have the oportunity to examine.

Nephrops norvegicus.

In Nephrops the structure and relations of the dotted substance is so quite similar to what they are in Homarus, that the above given description will suit in all particulars for both Homarus and Nephrops.

The Nereidæ.

Amongst the Polychætes, it is chiefly the Nereidæ (especially Nereis virens) that I have examined in respect of the structure of the dotted substance; I shall therefore in this chapter mention those animals only.

As the dotted substance of the ventral nerve-cord is the simplest, and easiest to investigate, we will confine ourselves to it.

In the ventral nerve-cord of Nereis the dotted substance has, as is well known, a situation and extension very different from what is the case in the ventral nerve-cord of Homarus. There are no distinct ganglia, and no distinctly separated longitudinal commissures, the dotted substance has, thus, a more uniform extension in two longitudinal rods along the whole nerve-cord. Its composition of tubes is, if possible, still more evident than it is in Homarus. In transverse sections of the nerve-cord we find, as before mentioned (vide fig. 14), a reticulation with small and large meshes. In longitudinal sections we find a longitudinal striation, which shows that the meshes of the reticulation, seen in transverse sections, are transsected thick and thin longitudinal nerve-tubes. The whole so-called dotted substance of the ventral nerve-cord of Nereis consists, thus, principally of longitudinally running tubes.

On close examination of sections under high powers of the microscope it will, however, be seen that slender tubes pass in all directions between these longitudinal tubes; this is very prominent in some small masses, which are especially ventrally situated on each side of the nerve-cord, and which have a more minute and granular appearance than the rest of the dotted substance.

In longitudinal section such masses will be found to occur especially near places where peripheral nerves issue. Fig. 66 represents such a place taken from a longitudinal section. *a a* are parts of the central longitudinal septum, dividing the ventral nerve-cord longitudinally into two lateral cords (cf. fig. 10, *a*). Smaller and larger nerve-tubes (*tt''*, *tt'*, *tt*) are seen passing the septum, from one side of the nerve-cord to the other, a great many of these small nerve-tubes (*tt'''*) seem to pass to the origin of the peripheral nerve, which origin lies to the left of the illustration. Some large nerve-tubes (*pt, pt*) pass to the same peripheral nerve. Amongst the longitudinal nerve-tubes smallish (*t*) as well as largish (*tt*) tubes may be seen. The nerve-tubes running transversally between these longitudinal nerve-tubes are, also, of different sizes. Besides nerve-tubes, neuroglia-fibres with transverse courses also occur, and they are often difficult to distinguish from nerve-tubes. Such transverse neuroglia-fibres may, for instance, be distinctly seen in the dotted substance on the left hand side of fig. 67.

Neuroglia-nuclei occur somewhat sparingly in the dotted substance of the ventral nerve-cord of *Nereis*. They are oblong, have a granular appearance, and are generally longitudinally situated in the sheaths of the longitudinal nerve-tubes (vide fig. 66, *n*). They are, however, also transversally situated in the sheaths of transverse nerve-tubes (vide fig. 66, *n'*).

To find the origin and course of the various tubes constituting the dotted substance has been very difficult, as I have not yet succeeded in obtaining any staining by the chromo-silver method in the nervous system of these animals. As far as I have been able to penetrate by help of the common staining with hæmatoxylin etc. (which has before been described) I believe, however, to have found that the constituents of the dotted substance have chiefly (if not wholly) the same origin in these animals as they have in Homarus.

I have found ganglion cells with nervous processes wich were directly transformed into nerve-tubes, but from which side-branchlets were given off (vide fig. 44). I have, however, not been able to trace such nerve-tubes (direct continuations of nervous processes) into peripheral nerves; those which I have observed, have partly had a longitudinal course in the nerve-cord.

I have also found ganglion cells with nervous processes, which subdivided in the dotted substance, and seemed to be entirely broken up into slender branches.

I have observed longitudinal nerve-tubes which seemed to subdivide and be broken up into branches, similar to those longitudinal nerve-tubes I have observed in the ventral ganglia of Homarus.

I have also observed side-branchlets to be given off from longitudinal nerve-tubes passing along the nerve-cord.

Finally, I have observed peripheral nerve-tubes springing from the dotted substance, being apparently formed by a union of slender tubes.

Peripheral nerve-tubes springing directly from ganglion cells, I have, as above mentioned, not been able to trace out, but I have been able to trace peripheral nerve-tubes for considerable distances through the dotted substance passing frequently over to the other side of the ventral nerve-cord, and I believe that they often spring directly from ganglion cells.

From the little we have seen, we may therefore conclude, that the constituents of the central mass, or dotted substance, of the ventral nerve-cord of Nereis are chiefly the same as those we have found in Homarus; the principal constituent is, however, in Nereis longitudinally running nerve-tubes, and the other constituents are comparatively sparingly present. Masses of a complicated web or plaiting of extremely slender tubes, similar to that which is described in the ganglia of Homarus, do not occur in Nereis.

Upon the whole, the dotted substance of Nereis is very simple in its structure, and represents a primary state.

The suggestion of E. ROHDE [1]) that the ventral nerve-cord of the Polychætes are only developed peripheral nerves because of the longitudinally running nerve-tubes, which should indicate the nerve-origin, is I think not very well founded. If that is right, the ventral nerve-cord of all invertebrated bilaterates which have a nerve-cord, must also spring from peripheral nerves only, because in all of them there are, according to my investigations, a great many longitudinal nerve-tubes, indeed the spinal nerve-cord of the vertebrates must also have the same origin. I do not know if this is ROHDE's opinion? According to this view the primary state of the nerve-system must be a central brain from which nerves only issue. Some of these nerves must then at a later stage, by an emigration from the brain, or in some other manner, have got ganglion cells.

---

[1]) l. c. 1886.

I do not think it necessary to discuss any further the many difficulties which would arise from such a theory; in my opinion, it is quite contrary to all known laws of evolution of the nervous-system. The longitudinal nerve-tubes have, in my opinion, another origin and significance, but this is not the place to enter into researches of this nature, we will defer this for a later occasion.[1])

---

[1]) Dr. *E. Rohde* describes »die Leydig'sche Punktsubstanz« of the Polychætes in the following way: »Untersucht man das Gehirn der Polychæten auf feinen Schnitten, so erkennt man, das dieselbe aus sehr vielen und feinen Fäserchen besteht, welche wirr durch einander ziehen und bald im Längsschnitt als Linien, bald im Querschnitt als Punkte erscheinen. Das Bauchmark hat im wesentlichen dieselbe Structur, nur überwiegen hier längsverlaufende Fäserchen, welche aber zahlreich von schiefen und queren gekreuzt werden. Querschnitte und Längsschnitte zeigen im Gegensatz zum Gehirn im Bauchstrang ein verschiedenes Bild, die Längsschnitte mehr Linien, die Querschnitte mehr Punkte.« It will be seen that this description by *Rohde* differs very much from mine. His »Fäserchen« are what I call sheaths of the nerve-tubes. It is strange, however, that he does not seem to have observed the reticulation which is produced in sections by the transsections of the nerve-tubes, and which is very prominent and distinct, for instance, in transverse sections of the ventral nerve-cord; he has perhaps applied unsuccessful staining-methods. In his description of the large nerve-tubes he mentions, however, fibrous sheaths enveloping them. Of one of the colossal nerve-fibres of *Sthenelais*, he says, that »it arises by a union of two nerve-fibres coming directly from two colossal ganglion cells in the brain, — and that it is enveloped by a fibrous sheath, which is at first closely applied to it, but in its further course separates from it and then encloses a cavity, which constantly becomes larger posteriorly and in the middle of the body attains an enormous diameter.« The nerve-fibre, which almost disappears in its wide sheath, gives off fine processes, traversing the whole cavity and apparently penetrating into the sheath. Towards the posterior extremity of the body the cavity becomes smaller, and conditions corresponding to those of the anterior extremity are reestablished.

As I have not examined *Sthenelais*, I can have no opinion of the correctness of this statement; I have not met with any structure similar to it in the Polychætes I have examined. From *Rohde*'s own description it appears to me that there must have been some irregularities in his preparations.

As to the origin of the large nerve-tubes, *Rohde* states that they spring directly from ganglion cells. As to their terminations, his opinion is that they are generally broken up into »feine Fäserchen«, wich in transverse sections are seen as minute dark points or dots.

The processes of the ganglion cells (which are »ausnahmslos unipolar«) pass, in his opinion, to the dotted substance partly »begleitet von Subcuticularfasern, welche aber bald nach ihrem Eintritt verschwinden.« What is meant by »Subcuticularfasern« I do not exactly understand, perhaps it is the neuroglia-sheaths surrounding the processes. The smaller processes »gehen direct in die centralen Fäserchen über.« The larger processes are partly enveloped by a sheath, being a continuation of the sheath of the ganglion cells. He believes it to be probable »dass sie durch pinselförmige Auflösung in die Fäserchensubstanz übergehen.«

Lumbricus agricola.

In Lumbricus, the dotted substance of the central nervous system represents a more developed state than it does in Nereis, it having a much more complicated structure; this is at once prominently apparent in transverse and longitudinal sections. The substance consists of tubes, and exhibits in sections a minute reticular appearance, the tubes being generally very slender; they vary a great deal as to their diameter, some are extremely slender, others are thicker but their average size is small. The tubes have, in the ventral nerve-cord, a longitudinal course, principally as may be seen in longitudinal sections. Between the longitudinal tubes, slender tubes are, however, interwoven, running in all directions, and forming a complicated web or plaiting. This plaiting is present to a much greater extent than is the case in Nereis, and at the same time the diameters of the tubes are generally much smaller than they are there.

As is well known there are no ganglia in the ventral nerve-cord of Lumbricus, and thus the dotted substance is uniformly extended along the whole nerve-cord.

The fibrillæ described by many previous authors, e. g. CLAPA-RÈDE, are, in my opinion, as before mentioned, only the transsected sheaths of the tubes which are the real constituents of the dotted substance.

In respect of the origin of these tubes forming the constituents of the dotted substance, it may be said so far as my experience goes, that their origin is quite similar to what is found in Nereis and Homarus.

As to the nervous processes of the ganglion cells, I have observed the same two types as are described in Homarus, viz. 1) nervous processes retaining their individuality and directly becoming nerve-tubes (either of a peripheral nerve or running longitudinally in the ventral nerve-cord [fig. 84]) but which give off side-branchlets to the dotted substance (fig. 71, *f*, *g*), and 2) nervous processes which lose their individuality and are, by subdivisions, entirely broken up into slender tubes losing themselves in the dotted substance (fig. 71, *a*, *d*, *h* and fig. 72).

As to the course of the longitudinal nerve-tubes of the dotted substance, and the double origin of the peripheral nerve-tubes, I have observed conditions very similar to those stated of Homarus, but my observations have been rather imperfect, as I have not obtained such staining in Lumbricus as I have in Homarus.

Neuroglia-nuclei occur sparingly in the dotted substance of Lumbricus (fig. 11, *k*; fig. 71, *n, n'*).

The Molluscs.

In *Patella* we find the dotted substance to have an appearance rather different from that of the animals before described. The elements of the dotted substance are, here, much smaller, and more difficult to trace out than those we have hitherto examined.

On a careful examination of successfully stained transverse sections through the pedal nerve-cord of *Patella*, it is possible to observe a minute reticulation with very small but distinct meshes. It is this reticulation which HALLER in the *Rhipidoglossa*, and RAWITZ in the *Acephales*, describe as nervous reticulation. Besides the meshes which have extremely slender walls, a great many dark dots are also seen. These dots are situated in the walls of the meshes, and chiefly in their corners where several walls of various meshes meet. It is these dots which RAWITZ has described as varicose thickenings in the nervous fibrillæ which he supposes to form this reticulation in the Acephales.

On examination of longitudinal sections through the pedal nerve-cord, it is seen that the dotted substance has, here, a somewhat different appearance. The reticulation seen in transverse sections is not present to such extent, on the other hand longitudinally running distinctly stained fibrillæ are very prevalent. These fibrillæ are extremely slender, and are stained in the same way as the reticulation in transverse sections. The intervals between those fibrillæ are very small and are of about the same size as the diameters of the meshes of the reticulation. This indicates that the meshes are transsected tubes, and that the longitudinal fibrillæ are partly the transsected sheaths of these tubes. It is consequently, in so far, a structure somewhat similar to what we have found in *Lumbricus* and *Nereis*. In longitudinal sections, but especially in oblique ones, it is seen that the dark dots visible in transverse sections are transsected fibrillæ, chiefly running along the concreting edges of the tubes. The question is, now, whether these fibrillæ belong to the neuroglia, and are only thickenings in the fibrous sheaths of the nerve-tubes, or whether they are real nerve-fibrillæ? To decide this question, I have examined fresh preparations as well as macerated ones, but I must admit, that in this respect I have not till yet succeeded in getting a clear idea of the real relations in the structure of the dotted substance of *Patella*. In fresh isolated preparations of the dotted substance plenty of fibres of various sizes are seen. The greater part of them are extremely slender. In fig. 85 some of these fibres from fresh preparations are illustrated. They are present in every preparation to unlimited extent, and do not seem to be formed in any artificial way, but only to be just isolated. In macerated preparations they are still more prominent (fig. 86). They may, here, be isolated, and traced for long distances through the dotted substance; to some extent, they are of a rather uniform size with a small diameter; they have a smooth aspect but may sometimes, though seldom, be seen to give off extremely fine lateral-fibrillæ (vide fig. 85, *b*, *e*; 86, *a*, *b*); where such fibrillæ issue small varicose thickenings are generally present. In thicker fibrillæ subdivisions occasionally happen (vide fig. 85, *d*). It is obvious that a great deal of the thickish fibres found in macerated preparations are real nerve-tubes, as they may be seen to be direct continuations of nervous processes from ganglion cells; these fibres have also, to a certain extent, an aspect like what is characteristic for nerve-tubes. To decide the nature of the extremely slender fibrillæ, which occur in such abundance, is, however, much more difficult; their aspect is undeniably very like that of common neuroglia-fibres or connective-tissue fibres; a great many of them are so thin that they, even under the highest powers of the microscope, appear like lines. As, however, we may find fibrillæ of every transition-stage from those finest ones up to these largish fibres, which unquestionably are nerve-tubes, and as we may see such fine fibrillæ to be given off from these nerve-tubes as side-branches, or being formed as a final-product by the breaking up of the latter, we feel ourselves forced to define these fibrillæ as being to a great extent real nerve-fibrillæ. These nerve-fibrillæ or *tubes*, as we should more properly call them, are certainly of a much smaller size than most of the tubes we have hitherto found in the dotted substance of other animals; this may, however, be accounted for from the smallness of all the nervous elements in Patella.

On the other hand, it is obvious that a great many fibrillæ seen in fresh and macerated preparations belong to the neuroglia. Such fibrillæ may often be seen to issue directly from neuroglia-cells, or united with neuroglia-nuclei (vide fig. 79—81; 73, *nn*; 82, *nc*; 83, *n*, *nc*; 86, *nc*) which occur abundantly in the dotted substance.[1]) These

---

[1]) The reason why *Haller* and *Rawitz* deny the occurence of neuroglia-nuclei is, I think, that they have not recognized their real nature and describe them as »Schaltzellen« which they believe to be multipolar ganglion cells. *H. Schultze*, *Walter*, and others have given somewhat similar descriptions.

neuroglia-fibrillæ resemble in their aspect the fine nerve-fibrillæ, so very much, that I am not at present in a position to point out any distinct difference between them.[1])

As to an interfibrillar substance, which many authors describe, it may be stated that, in fresh preparations, I have occasionally observed hyaline pearls adhering to the sides of the fibrillæ (vide fig. 85, *b*). In my opinion these pearls do not, however, spring from an interfibrillar substance, but are pearls of hyaloplasm, springing from destroyed nerve-tubes.

The origin of the constituents, forming the dotted substance, we will find to be quite correspondent to what is found in *Homarus*. All the nervous processes of the ganglion cells are, as already mentioned, directed towards, and penetrate into, the dotted substance.

As in *Homarus*, and the other animals examined, it may be seen that some *nervous processes* (vide fig. 73, *k*, *m"*; 74, *a*, *c*, *d*, *f*; 75; 76; 78; 82, *a*; 83, *a*) *retain their individuality*, in their course through the dotted subsance, and as far as they can be traced *are not seen to subdivide*; at intervals, however, *they give off some few fine side-branchlets*, at the origin of which small varicose thickenings frequently occur (fig. 74, *a*, *f*; fig. 75, 76).

Other *nervous processes* (vide fig. 73, *b*, *f*; fig. 74, *h*, *i*; fig. 77) *lose their individuality*; they *subdivide* and are *broken up into slender nerve-fibrillæ*.

We have thus two types of nervous processes, like what is found in *Homarus* etc.

As to the course and termination of the many longitudinal nerve-tubes, which form such a material part of the dotted substance of the pedal nerve-cord, little can be said at present. In some of them I have observed side-branchlets to be given off, in others I have occasionally seen subdivisions; but upon the whole my present methods of investigation have been insufficient to trace out these very slender structures. A great many of them are, however, easily seen to pass into peripheral nerves, forming peripheral nerve-tubes.

As to the origin of the peripheral nerve-tubes, my investigations are also in that respect very imperfect. I have observed nerve-tubes which, probably, come directly from ganglion cells, as they could be traced for long distances in their course through the dotted substance without subdividing, only giving off some few side-branchlets;

---

[1]) Vide also my description of the nerve-tubes of the Mollusca p. 94—96.

they were often directed towards groups of ganglion cells; I have not, however, succeeded in observing any direct combination with the latter. Other peripheral nerve-tubes could be seen to subdivide soon after their entrance into the dotted substance.

A significant relation is that, a great many nerve-tubes of the peripheral nerves are considerably larger in the peripheral parts of the latter, than they are near their origin in the pedal nerve-cord, and in the latter itself. As to the size of the nerve-tubes; the pedal nerve-cord and the first parts of the peripheral nerves are very much alike, and it is presumably the same substance, viz. dotted substance, that forms both. To make this correspondence quite complete, a great many ganglion cells also occur in the first parts of the peripheral nerves, especially in the larger ones (e. g. vide fig. 20, *gc*). Thus, it really looks as if the first parts of the larger peripheral nerves, issuing from the pedal nerve-cord of Patella, also belong to the central nervous system.

The fact that the nerve-tubes in the peripheral parts of the nerves have generally a larger diameter than the nerve-tubes near the origin of the nerves, can not, in my opinion, be explained in any other way than that the former to a great extent arises by a union of the latter. In other words, the roots of the nerves contains, to a great extent, only the elements of the peripheral nerve-tubes, and the formation of the latter does not take place in the pedal nerve-cord or the central nervous system only but also in the peripheral nerves themselves.

The Ascidians.

I have already, in a previous paper, given a preliminary report of the results of my investigations on the dotted substance of the Ascidia (vide l. c. 1886), and my present description will partly be, only a confirmation of what is there stated.

In well prepared transverse sections of the brain, the dotted substance exhibits a distinct reticulation with minute round meshes (vide fig. 87). These meshes are comparatively larger, and more distinct, than those we found in the dotted substance of Patella; their walls are distinct and rather uniform. Upon close examination of transverse as well as longitudinal sections, they may be seen to be produced by a transsection of nerve-tubes, and the walls of the meshes in the reticulation are only the transsected sheaths of the tubes. This may partly be seen in longitudinal sections where a great many nerve-tubes are longitudinally transsected and show a longitudinal course; it is especially prominent near the anterior and posterior extremities of the brain where the large anterior and posterior nerves issue, and where, consequently, a great many nerve-tubes converge towards their roots, causing longitudinal sections to have a longitudinal striation (fig. 88), whilst transverse sections of the same parts exhibit a distinct reticulation with meshes of about the same diameter as the intervals between the longitudinal lines of the longitudinal sections. In the mesial parts of the brain, the nerve-tubes run in more diversified directions, and thus longitudinal as well as transverse sections exhibit a reticulation, as tubes are transversally transsected in both; some tubes are, however, obliquely, or partly longitudinally transsected and appear then, in the sections, as oblong or elongated meshes. As the nerve-tubes forming the dotted substance vary in diameter, the meshes are consequently also of various sizes.

In the walls between the meshes, slender fibrillæ are seen to run in all directions, forming an intricate web or plaiting between the tubes, but neither these fibrillæ nor the larger nerve-tubes can be seen to anastomose with each other, and thus no real reticulation is formed by either of them; the *dotted substance consists of a web or plaiting of nerve-tubes and fibrillæ*[1]), and it is only the *transsection of the tubes which, in section, gives the substance the appearance of containing a reticulation.*

In macerated preparations of the dotted substance, a great many tubes and fibrillæ are easily isolated. Fig. 89 represents such a preparation. It is here distinctly seen that, no reticulation and no anastomoses are present. From some tubes side-branches are given off.

On examination of the origin of these nerve-tubes and fibrillæ it will be seen that they have the very same origin as they have in the animals previously examined. A great many come from ganglion cells, and a great many pass on to constitute the peripheral nerves, and from both fibrillæ and side-branches are given off.

The nervous processes issuing from the ganglion cells are of

---

[1]) These fibrillæ are partly nerve-fibrillæ, partly neuroglia-fibrillæ. The nerve-fibrillæ have, probably all of them, a tube-structure like the larger nerve-tubes; i. e. an external sheath enclosing a semi-fluid hyaline contents, hyaloplasm. In nerve-fibrillæ which are not too slender, this tube-structure may be seen on close examination. The neuroglia-fibrillæ are often seen issuing from neuroglia-cells which frequently occur in the dotted substance. I am not in a position to say anything with certainty about their structure at present.

the same two types previously described. Some processes retain their individuality, and pass directly to form nerve-tubes, whilst giving of slender side-branches (fig. 55—57; 87, c; 89, ge), other processes lose their individuality and are broken up into slender branches losing themselves in the plaiting of the dotted substance (fig. 87, a, b, d, e). I have been able to trace both these types of processes, only for comparatively short distances through the dotted substance, but still the difference between them is generally possible to distinguish.[1])

As before said, a great many of the nerve-tubes of the dotted substance go to constitute the peripheral nerves. Many of these nerve-tubes are, consequently, direct continuations of the nervous processes from ganglion cells of the one type; these nerve-tubes probably form peripheral nerve-tubes directly. A great many of the other nerve-tubes and fibrillæ, constituting the nerves, are either continuations of the slender branches arising from the subdivisions of the nervous processes of the other type (these fibrillæ are extremely slender, and in the roots of the nerves they unite to form thicker tubes), or they spring from a union of similar fibrillæ or branches; perhaps the side-branches given off from the nervous processes of the first type also contribute to the forming of these nerve-tubes.

As constituents of the peripheral nerves, present in their roots, we have thus elements of two kinds; viz. 1) nerve-tubes which are direct continuations of nervous processes, but which, however, are partly strengthened by the junctions of slender side-branches, and 2) nerve-tubes arising from a union of slender tubes or fibrillæ.[2])

---

[1]) As is mentioned in my previous paper on the nervous system of the Ascidia (l. c. 1886), small ganglion cells occur in the dotted substance (vide fig. 87, f, g). Those cells have generally a tripolar or multipolar form (bipolar cells also occur). Fig. 57 represents such a cell, isolated, in a macerated preparation. The protoplasmic processes of these cells are, generally, directed peripherically towards the external layers of the brain, whilst the nervous process has very often a longitudinal course through the dotted substance. I have often, both in macerated preparations and in sections, been able to trace such nervous processes for some distance of their course through the brain, but have rarely observed any subdivision of them. I think it therefore probable that they, to a certain extent, are directly continued into peripheral nerve tubes; some of them are, however, broken up into branches losing themselves in the dotted substance.

[2]) It may here be mentioned that, I have observed ganglion cells situated in the roots of the nerves and sending their nervous processes directly into the nerve. In fig. 88, which represents a part of a longitudinal section through the anterior end of the brain of a *Corella*, such a ganglion cell (gc) is seen situated at the root of one of the anterior nerves. I am not in a position to state whether the nervous processes of these cells send off side-branches.

Neuroglia-nuclei are scattered through the dotted substance of the Ascidia; they distinguish themselves from the nuclei of the before mentioned ganglion cells, also occurring in the dotted substance, by a darker staining, a more oblong form, and a more granular appearance, they are also generally of a smaller size than the latter. As before said, neuroglia-fibres are very often seen in connection with these neuroglia-nuclei. Sometimes neuroglia-nuclei are seen adhering to the nervous processes issuing from the ganglion cells (vide fig. 87).

Summary.

If we now gather together the results of our previous researches on the structure of the dotted substance, and if we assume them to be applicable to the dotted substance of all invertebrated bilaterates — which we may probably do, seeing the correspondence between the principal characteritics in the structure of the various groups investigated — we may give the following summary of our investigations:

The dotted substance of all *invertebrated bilaterates* consists chiefly of nerve-tubes, and primitive-tubes (and nerve-fibrillæ which are only small primitive-tubes); these tubes consist of a neuroglia-sheath, and a semi-fluid contents (hyaloplasm), they have consequently a structure similar to the primitive-tubes of the nerve-tubes, only that their sheaths are stronger than the spongioplasmic ones of the latter.

The *tubes and fibrillæ forming the dotted substance do not anastomose with each other*, but form, only, a more or less intricate web or plaiting. The reticulation seen in sections, and described by various authors as a real nervous reticulation, is no reticulation, but is *produced by the transsection of the tubes* forming the dotted substance, and *the meshes of the reticulation are only the transsected sheaths of these tubes*. The „*interfibrillar substance*", described by various authors, is the hyaline hyaloplasm, which forms the contents of the tubes, and is the real nervous substance.[1]

---

[1] I think it probable that it is the same substance which *Rawitz* describes as myeloid substance. *Hermann*'s ›Zwischensubstanz‹ (cf. p. 46) is evidently hyaloplasm. The ›Knotenpunkte‹, described by Hermann, are, I suppose, thickenings in the sheaths of the tubes, they can not be neuroglia-nuclei, as they seem to occur in too great abundance.

The origin of the nerve-tubes and nerve-fibrillæ (primitive tubes) of which the dotted substance consists is the following:

1) The branches of those nervous processes which lose their individuality and are entirely broken up into slender branches.

2) The side-branches of those nervous processes which do not lose their individuality but directly become nerve-tubes whilst giving off side-branches on their course through the dotted substance.

3) Those longitudinal nerve-tubes which run along, through the dotted substance, especially of the ventral nerve-cord, and which partly pass into peripheral nerves.

4) The side-branches given off from those nerve-tubes.

5) The branches of those longitudinal nerve-tubes which are entirely broken up into slender branches losing themselves in the dotted substance.

6) The slender tubes or fibrillæ which unite to form those peripheral nerve-tubes which exclusively spring from the dotted substance.

7) The side-branches joining those peripheral nerve-tubes which spring directly from ganglion cells, i. e. which are direct continuations of nervous processes from ganglion cells. These side-branches are consequently partly the same as those mentioned in 2.

Besides nerve-tubes and fibrillæ, *neuroglia-cells* and *fibres* are also present in the dotted substance of all invertebrated bilaterates. The neuroglia-nuclei have generally an oblong shape with a granular appearance.

---

4. The combination of the ganglion cells with each other, and the function of the protoplasmic processes.

In the historical introduction to this paper it is pointed out that two views as to the combination of the ganglion cells have especially been prevalent amongst the previous authors (vide p. 64—65). We will now examine the conclusions in this respect our present researches must lead us to.

A direct combination between the ganglion cells is, as we have seen, *not acceptable.* In spite of the most persevering investigations I have not been able to find any direct anastomosis of indubitable nature between the processes of the ganglion cells. Where I thought to have found an anastomosis it always on application of the strongest lenses resolved itself into an optical illusion. In a few cases I have certainly observed short bridges extending between neighbouring cells, these bridges have, however, in my opinion, evidently been remnants of cell-divisions, and are therefore of but little interest to our present subject; they had not the appearance of common processes.

If a direct combination is the common mode of combination between the cells, as most authors suppose, direct anastomoses between their processes ought, of course, to be quite common. When one has examined so many preparations (stained by the most perfect methods) as I have, without finding one anastomosis of indubitable nature, I think one must be entitled to say, that *direct anastomosis between the processes of the ganglion cells does not exist, as a rule.* What previous writers have supposed to be anastomoses is, in my opinion, probably the neuroglia-reticulation generally extending between the ganglion cells, and the fibres of which are often difficult to distinguish from the processes of the latter.

Another objection against a direct combination, and which does not seem to have been thought of by a great many authors, is the existence of unipolar ganglion cells. How is it possible to explain an existence of unipolar cells, when we believe in a direct combination by anastomosing processes? Certainly we will be obliged to say with VIRCHOW, and others, that — the closer we examine the nervous systems — the fewer unipolar cells do we find; in other words we can not believe in the usual existence of unipolar cells.[1]

---

[1] A good instance of the results to which the common theory of the function and combination of the ganglion cells must bring us is a recent paper by *Fritsch* (op. cit. Arch. mikr. Anat. 1886). This author says, that the apolar ganglion cells have »lange genug in unserer Litteratur gespukt.« Improved methods of investigations have taught us that they were creation of our imagination. And he continues (l. c. p. 24): »Wenn ich jetzt nach reiflicher Ueberlegung erkläre, dass die unipolaren Ganglienzellen denselben Weg wie die apolaren wandern werden, so muss ich fürchten, augenblicklich noch dem energischsten Widerspruch zu begegnen.« I agree with the author in the latter point, he will certainly meet opposition. Further on in the same paper he says (p. 29): »dass es der Natur der Ganglienzellen überhaupt widerspricht, als einzelnes Element nur eine Verbindung mit der Peripherie zu haben.« And: »Eine wirchlich unipolare Zelle ist für den Organismus nicht viel mehr werth als eine apolare Zelle.« This is in my opinion a quite logic conclusion if we suppose that the common view of the nature of the ganglion cells is correct. When now, however, unipolar ganglion cells actually exist, can that easily be supposed to be the fault of the ganglion cells? or is there not a possibility that our view of the nature of the ganglion cells is incorrect? We can not change the the reality according to our ideas but we can change our ideas

But we must believe in it. We need, for instance, only go to the nervous systems of the lobster, and we shall be quite convinced of the existence of real unipolar cells. If those large ganglion cells of the lobster have several processes, it is strange that we should not be able to observe one of them by our present modes of investigation.

We are thus obliged to reckon with real unipolar cells, indeed we are obliged to reckon with nervous systems containing almost entirely unipolar ganglion cells, and how could direct combination between the cells be present where there are no processes to produce the combination. We thus see that also for that reason we are obliged to abandon the theory of the direct combination between the ganglion cells.

Two questions which will now present themselves are:

How is the combination between the ganglion cells produced? and what is the function of those processes which were previously thought to produce the direct combination between the cells?

We will first examine the latter question. As before described, the processes of the ganglion cells are of two kinds: *nervous* processes and *protoplasmic* processes. Of nervous processes each cell (unipolar or multipolar) has always one and never more; the nervous processes are always directed towards the dotted substance, or in a few cases they may pass directly into peripheral nerves (cf. p. 143, note 1). It was not, however, these processes which were generally supposed to produce the combination between the cells, but it was the protoplasmic processes.

The protoplasmic processes. — When a ganglion cell is bipolar or multipolar, then the processes it posesses, besides the nervous process, are protoplasmic processes. These protoplasmic processes are not directed towards the dotted substance, but generally have a peripheral direction towards the external layers of the central nervous system. In a great many cases I have been able to trace such processes just to their communication with the peri-

---

according to the reality, some people can, at all events; as the mountain would not walk to Muhamed, Muhamed had to walk to the mountain. In chapter 7 we will have an opportunity to discuss this subject somewhat more circumstantially.

(Regarding unipolar cells etc. vide also my quotation from Rawitz's paper p. 66.)

As subsequently may be seen, it is, in my opinion, of no importance as to the combination of the cells whether they are unipolar or not. I do not at all doubt the existence of the slender processes *Fritsch* describes in his ganglion cells from the *ganglion gasseri* of *Lophius*, but I suppose them to be protoplasmic processes, i. e. *nutritive* processes.

neurium enveloping the nervous system (e. g. vide fig. 50, *ppr*). In Patella I have even found ganglion cells with similar processes in peripheral nerves (e. g. vide fig. 20, *gc*, the nervous process of this cell is not distinctly seen in the illustration). Whether the protoplasmic processes, as a rule, extend to the perineurium, like what is the case in the spinal cord of *Myxine* (as will be subsequently described) I do not know, as they are generally lost in the neuroglia reticulation extending between the ganglion cells (e. g. vide fig. 47, 50, 71, 73). In the brain of the *Ascidia*, as well as in the nervous system of *Patella*, I have found bipolar and multipolar ganglion cells situated in the dotted substance, and sending their protoplasmic processes towards the periphery of the nervous system

Like Prof. GOLGI I belive the function of the protoplasmic processes to be *a nutritive one;* when the ganglion cells can not get sufficient nutrition in their neigbourhood, they have to send processes towards the periphery of the nervous system, or out into the loose neuroglia reticulation, where there is sufficient nutritive fluid for the processes to absorb.[1]) This is the reason why the protoplasmic processes have generally a peripheral direction, and why they chiefly occur in cells which are somewhat deeply situated in the nervous system, i. e. at some distance from the periphery. The ganglion cells situated near the periphery do not need any special organs to absorb their nutrition from the fluid surrounding them, neither do the ganglion cells situated in a loose neuroglia-reticulation need any (compare the ganglion cells of the lobster).

Though we have of course no proof of the real function of the protoplasmic processes, I think this theory of their nutritive function is very probable, and explains a great deal which, else, may seem inexplicable. What other function could we suppose them to have when they terminate under the perineurium, what they actually do, at all events, in a great many cases? — Indeed, I do not, at present, see any acceptable explanation besides this one suggested by GOLGI. Subsequently, we will have an opportunity to state some other facts which will still further add to its probability

The combination of the ganglion cells. — Having thus elucidated that no combination between the cells can be produced by the protoplasmic processes, and as it is very improbable that any combination between them can be produced by the neuroglia, I suppose it to be most probable, if not certain, that if any

---

[1]) Where there are blood-vessels in the central nerve-system (e. g. in the Vertebrata) the protoplasmic processes can often be seen to be directed towards the connective tissue surrounding the blood-vessels.

combination upon the whole exists between the ganglion cells, then this combination must be produced by the nervous processes. That such a combination can exist, with the dotted substance as a medium, we will easily understand when we think of the course of the nervous processes. As is previously mentioned, there may be drawn a distinction between *two types of ganglion cells* in respect of the course of their nervous processes; viz. 1) ganglion cells with nervous processes which *directly become nerve-tubes* and thus do not lose their individuality, though they have no isolated course *but give off side-branches to the dotted substance*; 2) ganglion cells with nervous processes which lose their individuality and by subdivisions are *entirely broken up into slender branches*, losing themselves in the dotted substance.

We have, thus, slender branches given off to the dotted substance, by both kinds of nervous processes. What is to prevent these branches standing in communication with each other, seeing that they are so intimately intermixed. Certainly, I have not been able to observe such a communication, but seeing how very intricate these structures are, we need scarcely at present expect to do so. In fact, I do not see any objection which can justly be made to a combination existing between the ganglion cells by help of the nervous processes and their branches. Still, I do not lay much stress upon such a combination existing, or not existing, as in a subsequent chapter will be explained; the principal question, in my opinion, is, how is the *combination between the nerve-tubes* produced.

5. The nervous elements of Amphioxus and Myxine.

Though I propose to reserve the nervous systems of *Amphioxus* and *Myxine* for a future special memoir, I will, here, for the sake of comparison, make some remarks on the structure of their nervous elements.

Amphioxus lanceolatus.

As is known from the descriptions of previous writers [1] the spinal cord of Amphioxus has, along the centre, a groove or canal extending from the dorsal side and surrounded by and epi-

---

[1] *Owsjannikow* (l. c. 1868) and *Stieda* (l. c. 1873) may especially be mentioned. I regret to say that I have had no access to *Stieda*'s paper; I only know it therefore from the abstract of it in Hofmann's and Schwalbe's Jahresb. d. Anat. u. Phys. Bd. 2. 1875.

thelium of cylindrical cells, and outside this epithelium a limited number of ganglion cells (vide fig. 90). The rest of the spinal cord consists of white substance.

The white substance and the nerve-tubes. — As is easily seen in transverse sections, the white substance of Amphioxus consists, chiefly, of longitudinal (partly also transverse) nerve-tubes, which in transverse sections are transsected and give the section the appearance of containing a reticulation (vide fig. 90). Some of the longitudinal nerve-tubes are very large, and very much resemble, in their appearance, the large nerve-tubes which are, for instance, present in the ventral nerve-cord of *Nereis*. They are especially situated in the lateral ventral parts of the white substance (vide fig. 90). The large tubes, which OWSJANNIKOW has described as blood-vessels, are colossal nerve-tubes. There is one colossal nerve-tube running on the ventral side of the nerve-cord, just under the central canal (fig. 90, *vnt*). One colossal nerve-tube generally runs, laterally, in each side of the spinal cord (fig. 90, *lnt*). These three nerve-tubes, especially the ventral one, are so far as my experiece goes, the largest ones existing in the nervous system of Amphioxus.

The small nerve-tubes vary very much in size, and some of them have an extremely small diameter.

There are no blood-vessels present in the spinal nerve-cord.

The white substance is traversed by a great many radiating fibres, issuing from the cylindrical cells of the epithelium investing the central groove, and passing to the sheath enveloping the spinal cord, with which sheath they unite.

The structure of the nerve-tubes. — The nerve-tubes have a structure, quite similar to what is found in the invertebrates we have before examined. They consist of a rather deeply staining *sheath* and a less staining *contents*. The *sheaths* are generally very thin. The *contents* consist of *primitive tubes*, quite similar to those which are, for instance, present in the nerve-tubes of Homarus. The primitive tubes may easily be seen in the large nerve-tubes (vide fig. 91); the spongioplasmic walls separating the tubes containing hyaloplasm are here very distinct.

The ganglion cells. — The ganglion cells occur rather sparingly. They are situated on both sides of the central groove outside the epithelium. Some ganglion cells are very large (vide fig. 90, *gc*) but generally they have a moderate size. They are usually multipolar, have always only a single *nervous process*, the other processes are *protoplasmic* ones. The protoplasmic processes seem generally to traverse the white substance and to extend to the envelope of the spinal cord. In a great many cases, I have been able to trace them to the external layers of the white substance (vide fig. 90, *pp*; 92, *pp*). In some cases I even believe to have observed cells which send protoplasmic processes to the envelope on both sides of the spinal cord; fig. 90, *gc* represents such a case; though there is not quite complete continuation of the process in this section, it seems, however, to be the same process which crosses the central canal, subdividing and sending its branches just to the external envelope on the other side. The relations of the protoplasmic processes seem, thus, to be the same in Amphioxus as are found in the other animals, examined, where such processes were present; viz. they are peripherically directed (here traversing the white substance). If we assume that they have a nutritive function this is easily understood, because as there are no blood vessels present within the nerve-cord, the processes must penetrate to the external layers to absorb nutrition.

I have not yet had sufficient opportunity to investigate the course of the nervous processes, and will therefore reserve them entirely for the future memoir.

The protoplasm of the ganglion cells has, in sections, a reticular appearance, produced, as I suppose, principally by a transsection of primitive tubes, composing the chief constituents of the protoplasm (vide fig. 92, *gc*). These tubes penetrate into both the protoplasmic and the nervous processes and form their contents, whilst giving them a longitudinal striation. The protoplasmic processes have, however, a more granular appearance than the nervous processes.

The cylindrical cells of the epithelium investing the central groove (or canal) have, in their external extremities, processes which traverse the white substance, and unite with the envelope of the nerve-cord, as already mentioned. These fibres are partly united into bundles (vide fig. 90), which apparently divide the longitudinal nerve-tubes into various columns. This division into columns is, however, only partial as the bundles do not form any continuous septa; they are only seen in some sections, whilst in other sections bundles of similar fibres may occur on other places. There are, however, some definite places where such bundles of fibres especially occur; as some of the most prominent of these, two places, one on each side of the ventral colossal nerve-tube, may be mentioned (fig. 90, *ns, ns*).

I have often observed very strong fibres or even bundles of

fibres of this kind directed towards the nerve-roots, especially the dorsal ones, and penetrating into the nerves.

It is obvious that the fibres, radiating from the epithelial cells of the central groove, and running between the longitudinal nerve-cord, have just the same significance as the neuroglia-fibres issuing from the neuroglia-cells of higher vertebrates. No such neuroglia-cells are found in Amphioxus as are, for instance, found in *Myxine* (vide sequel). We may, therefore, assume that, *these epithelial cells are the real neuroglia-cells of Amphioxus*, and that *this neuroglia represent the most primary state found amongst the Vertebrata*. In the description of the neuroglia of Myxine we will have an opportunity to return to this subject.

Fibres are often seen crossing the central groove; these fibres seem partly to be neuroglia-fibres, partly nerve-tubes and, as we have seen, partly also processes from ganglion cells.

As to the origin of the nerve-tubes of the ventral and dorsal nerve-roots, my investigations are yet in this respect very incomplete, and must be reserved for the furture paper.

Myxine glutinosa.

It is only the nervous elements of the spinal cord we will, here, describe.

The white substance. — On examination of transverse and longitudinal sections of the spinal nerve-cord of *Myxine* it is easily seen that the white substance chiefly consists of longitudinal nerve-tubes, which vary a great deal in size. The large tubes, the so-called »MÜLLER's nerve-fibres«, are especially situated on the ventral side. Between the large and largish tubes and through the whole white substance a multitude of very slender longitudinal nerve-tubes occur; the white substance has thus in well preserved transverse sections a reticular appearance (vide fig. 100) quite similar to that of a section through the dotted substance of the ventral nerve-cord of an invertebrate (e. g. Homarus and *Nereis*). Between each of these tubes a multitude of extremely slender tubes (neuroglia-fibres and branches from the protoplasmic processes of the ganglion cells) run transversally. These transverse fibres are not very distinctly seen in so well preserved sections as fig. 100; they are, however, very conspicuous in sections of preparations where the contents of the longitudinal tubes have shrunk in so much that the transsected tubes have the appearance of empty vacuoles, the white substance has here become very transparent (vide fig. 101). They run from the grey substance and just to the periphery of the white substance (to the sheath of connective tissue). They seem to compose a significant constituant of the white substance. Subsequently they will be more circumstantially described.

The structure of the nerve-tubes is the same, in Myxine, as in the other animals examined. The nerve-tubes consist of a distinct external sheath, and a contents containing spongioplasm and hyaloplasm. The external sheaths are rather thin (vide fig. 100).

The contents of the nerve-tubes consist of *primitive tubes*. In the large nerve-tubes (MÜLLER's nerve-tubes), the primitive tubes are very distinctly seen (vide fig. 100, *Int*).[1]) Upon the whole, it is striking how very much these MÜLLER's nerve-tubes resemble the large nerve-tubes of the lobster; in Myxine I have even observed a tendency to the formation of an axis along the centre of some larg tubes (vide fig. 100, *Int'*) somewhat similar to what is described of the large nerve-tubes of the lobster (vide p. 87).

Medullated nerve-tubes are *not observed* in the spinal cord of Myxine.

The ganglion cells are situated in the grey substance; they are, as a rule, multipolar, have always one nervous process each, and generally several protoplasmic ones.

The protoplasm of the ganglion cells has, in sections, always a minute reticular appearance, with spongioplasmic meshes containing a less-staining substance; the meshes are, to a great extent, produ-

---

[1]) I think it is very strange that previous authors have not seen the composition of primitive tubes in the large nerve-tubes of the spinal cord of the fishes. In his description of the large nerve-tubes in medulla oblongata of *Lophius*, *Fritsch* says (l. c. 1886, p. 22): »Der Faserlängsschnitt zeigt, dass die Fibrillen nicht regelmässig parallel angeordnet sind, sondern sich vielfach durchflechten....« It is obvious, in my opinion, that *Fritsch* has seen the spongioplasmic walls of the transsected primitive tubes and has called them »Fibrillen«

Stilling has long time ago (l. c. 1855) decribed the contents of the nerve-tubes of the Vertebrata as consisting of slender tubes; he said himself in his paper that there would likely go a long time before anybody could state the correctness of his observations, as it was only by help of new and very high powers of the microscope he had been able to obtain his results. Indeed, there has gone a long time; no other writer has since been able to see *Stilling*'s »Elementar-nervenröhrchen«. His observations were made too early. They were of cource, imperfect, as he had not the same lenses and modes of investigation as we have at precent; a great deal of what he has seen, has been artificial products (e. g. his description of the myeline of the medullated nerve-tubes) but he has said that the *cylinder-axis consist of slender „Elementarrörchen"* and in this point I do entirely agree with him.

ced by a transsections of the primitive tubes of which the protoplasm chiefly consists. These primitive tubes extend into the processes and give them a longitudinally striated appearance, the dark lines of the striation being the spongioplasmic walls of the primitive tubes.

In sections of a great many, especially large, ganglion cells, there are in the protoplasm circular or oblong lightly staining areas resembling vacuoles (vide fig. 96, v) which are very similar to those, previously mentioned, in the ganglion cells of Homarus (cf. p. 103 & fig. 25, 27). These areas are transsected tubes of a lightly staining substance, probably bundles of primitive tubes, which circulate in the protoplasm and run into the processes. They are often obliquely or longitudinally transsected and produce then areas of an elongated shape (vide fig. 96, v'). I have observed very similar tubes or bundles in the cells of the spinal ganglia of *Myxine*, as will be subsequently described, and I have also observed similar structures in the ganglion cells of higher vertebrates *(Mammalia)*. It seems thus to be a relation of rather general nature, and it is therefore, in my opinion, strange that it has not previously been described. We know of course yet too little of these structures to attempt to say anything of their significance.

The sheath enveloping the ganglion cells. — The ganglion cells are always enveloped by a membran or sheath (vide fig. 96, *cm*), which also extends into the process and forms their sheaths (cf. fig. 96). This sheath is, I think, a product of the *neuroglia*, or has, at all events, the same origin as the latter. I have not, however, hitherto been able to observe any nuclei situated in the sheath or adhering to it. Between this sheath and the protoplasm of the ganglion cells there is, generally, a cavity[1], (fig. 96, *a*) filled with a loose, apparently reticular substance, which is extremy lightly staining; this substance is somewhat delicate, and is seen in well preserved sections only.

The protoplasmic processes of the ganglion cells have always a more or less peripheral direction; they subdivide, generally dichotomically a great many times, and the numerous extremely slender, branches resulting from these divisions run, between the longitudinal nerve-tubes of the white substance, towards the periphery of the spinal cord, where they terminate in small thickenings,

---

[1] It is probably the same cavities which are described by *Key* and *Retzius* (l. c. 1876) and which these authors supposes to be filled with lymphoid fluid.

or plates, just inside the sheath. The only methods which affords satisfactory staining of these processes is the chromo-silver method, but by its use I have been able to trace them to their extreme terminations (vide fig. 94, *ppr*; 104—105). I have, however, also in preparations stained with *borax-carmine* (fig. 93) or stained according to HEIDENHAIN's hæmatoxylin method (fig. 95) been able to trace the protoplasmic processes with their branches through the white substance and very close towards its periphery if not entirely to the external sheath.

The same ganglion cell may have protoplasmic processes, the various branches of which penetrate to the periphery, both on the ventral side and on the dorsal side of the nerve-cord (vide fig. 93, 95, 105, 106).

The slender protoplasmic processes and their branches resemble the processes of the neuroglia-cells very much in their aspect and it is indeed extremely difficult to draw any distinction between them. Both have the same appearance, both come from the grey substance, and both traverse the white substance towards the periphery where both as a rule terminate under the sheath enveloping the spinal cord. In fig. 109, a protoplasmic processes issues from the small ganglion cell *gc*, but this process can not be distinguished from the many neuroglia-fibres, *f*, in any respect but that it issues from a ganglion cell.[1]) A difference is sometimes that the neuroglia-fibres have a straighter course more directly towards the periphery of the white substance than the protoplasmic processes have. I have found no constant and reliable difference in the chromo-silver staining of these processes similar to what GOLGI has stated as regards the neuroglia cells and protoplasmic processes of the *Mammalia*, and what I also have had opportunity to observe myself in higher Vertebrata. Sometimes there was, however, also in *Myxine* a tendency towards a more reddish staining of the processes of the neuroglia cells.

The branches of the protoplasmic processes of one cell do not anastomose, neither do the processes from various cells anastomose. Though I have examined a considerable number of preparations, and although they have been most perfectly stained, in this respect, I have not yet observed a single case of indubitable anastomosis between protoplasmic processes.[2]) I believe, thus, that I am entitled

---

[1]) Vide also fig. 93, gc_2.

[2]) In my previous paper on Myxine (l. c. 1886) I have mentioned one case from the brain where I was in doubt whether there was an anastomosis or not. The more I have examined this preparation, the more do I doubt that an

to affirm that a *direct combination between the ganglion cells, by direct anastomosis of the protoplasmic processes does not exist.*

A thing which I take to be rather puzzling is, that processes which in their aspect can not be distinguished from common protoplasmic processes, are often seen to pass towards the periphery of the nerve-cord; arrived there, they do not, however, terminate, but run along the periphery and partly back into the white substance (vide fig. 102, *pr*, *br$_2$*) I have not found the terminations of such processes. Some other processes, with quite similar aspect, may be traced for very long distances, even crossing the ventral commissures and passing over into the other side of the spinal nerve-cord (vide fig. 102, *br$_4$*). A great many processes which I have observed to have a similar course are, near their origin in the cell, very thick and large, and they give off branches which terminate in the periphery, under the sheath, quite in the same way as common protoplasmic processes, whilst at the same time, the main stem of the process runs along, and takes a course as above described (vide fig. 93, *mp* and 94, *mp*). It has been possible to trace similar processes not only in chromo-silver stained preparations but I have also traced them for considerable distances in preparations stained with carmine or HEIDENHAIN's hæmatoxylin only (vide fig. 93 and 95). If, now, these processes are protoplasmic processes, and if the protoplasmic processes have a nutritive function, which makes it necessary for them to terminate under the sheath enveloping the spinal cord, it is, indeed, very difficult to understand why they pass through the nervous system for so long distances without terminating; why they pass to the other side; and why they do not terminate, but return again, when they are in the periphery? These are questions of so serious a nature that they oblige us to assume that either these processes are not wholly protoplasmic processes though they have their aspect, or that the function of the protoplasmic processes is not only a nutritive one. I will not decide between these alternatives at present, though I find the former one most probable. It can not be doubted, in my opinion, that some

---

anastomosis is really present. If there is an anastomosis, however, it may be accounted for as a remnant of a cell-division not yet finished, as the bridge between the cells is very thick and short.

Mr. *W. S. Colman* (l. c. 1884) has actually observed anastomoses of ganglion cells in the spinal cord of a human foetus; but there is, in my opinion, little doubt but that they are results of divisions of ganglion cells; in two of the cells there were even two nuclei, what obviously indicates that the cells are dividing.

of the branches given off from these processes are protoplasmic branches, as they have the same appearance and course and *actually terminate under the external sheath* in a way entirely similar to that of common protoplasmic processes (as also neuroglia-fibres, vide sequel). Could it be possible that these processes are mixed protoplasmic nervous processes? GOLGI has described and illustrated nervous processes issuing from protoplasmic processes near their origin in the cell (vide l. c. 1885). It may be, that these structures in *Myxine* are processes of a still more mixed nature. More complete observations on the course and termination of the main branch of these processes are necessary to decide in this question. I can not omit to state that it has not been possible to find any other processes which looked like nervous processes and which issued from the same cells as the »mixed processes«, described above.

The nervous processes. — My observations on the nervous processes are yet very scanty. The chromo-silver staining is the only one which can tell us anything, with certainty about the course of these processes, and upto this time it has been exceedingly difficult to get any perfect staining of them. They have a smooth and less granular aspect than the protoplasmic process. The slender branches have generally at certain intervals varioceles (vide fig. 103, 104).

From what little I have seen it is probable that in Myxine there are also two forms of nervous processes, viz.:

1) Nervous processes, which do not lose their individuality, as they are directly transformed into nerve-tubes; they have, however, no isolated course, as they give off slender side-branches. Fig. 103, gc, represents a ganglion cell with such a nervous process. The process has often varicose thickenings at the origin of the side-branches. Processes which have no side-branches, but sometimes varicose thickenings at certain distances, are also seen (vide fig. 94, gc_1, gc_2), I think, however, that the absence of side-branches issuing from the thickenings is only owing to imperfect staining. Nervous processes of this kind are often seen to pass through one of the commissures and over into the other side of the spinal cord (fig. 94).[1]

2) *Nervous processes* which lose their individuality and are entirely broken up into slender branches. Fig. 103, gc_3 and 104 represent ganglion cells with such processes. In their slender parts the branches

---

[1] In fig. 103, gc_2 is seen a ganglion cell with a process which crosses the commissure and which could be traced for some distance over into the other side; this process had the aspect characteristic for protoplasmic processes, but judging from its course I think it is a nervous process. At a a branch was given off.

of these processes have, generally, varicose thickenings at each places where they subdivide. They have, generally, also a great many varioceles where no side-branches are seen, I suppose, however, this is only owing to an imperfect staining, and that side-branches are given off at each variocele.

The branches of the nervous processes do not anastomose. — They contribute to form a reticulation, or rather interlacing of nervous fibrillæ which extends through the grey substance, and which also seems to extend into the white substance.

The origin of the peripheral nerve-tubes. — In horisontal sections of successfull chromo-silver stained preparations of the spinal cord, it is easy to observe that, *the nerve-tubes of the dorsal (posterior) nerve-roots*[1]) *subdivide dichotomically, just after their entrance into the spinal cord;* some of them subdivide even before they have reached the cord (vide fig. 111, a).

As far as my experience goes I dare to say that all nerve-tubes subdivide, it has in my opinion therefore, only been optical illusions which have occasioned previous outhors (e. g. RANSOM & THOMPSON [l. c. 1886] and others) to state that, they have observed some of these nerve-tubes to originate directly from ganglion cells; these authors will, I hope, forgive me this denial of their statements, when I tell them that such an illusion has once deceived myself, before I obtained perfect chromo-silver stained prepations (cf. my previous paper l. c. 1886). The fibres which I at that time observed (and which probably the other authors have also seen) were neuroglia-fibres issuing from neuroglia-cells in the grey substance and intermixing with the nerve-tubes of the dorsal nerve-roots (vide fig. 93, f_1.)

The two branches resulting from the dichotomical subdivisions of the nerve-tubes separate, and run in opposite directions, longitudinally, along the spinal cord. I have been able to trace these branches for considerable distances along the cord but, as a rule, I have found no new subdivisions; neither have I seen them passing over to the other side of the spinal cord. In a very few cases, I belive to have seen very slender side-branches to be given off (vide fig. 112, *br*). My investigations on this subject are very far from being finished, and I hope yet to be able to make important observations in this respect. From what is already observed,

---

[1]) It may be noticed that in my previous paper (l. c. 1886) there is in the English translation (Ann. mag. nat. hist. 1886) printed »dorsal nerve-rods« and »ventral nerve-rods« instead of *dorsal nerve-roots* and *ventral nerve-roots*.

I am, however, at all events, entitled to say that, the *nerve-tubes of the dorsal (posterior) nerve-roots have no direct origin from ganglion cells.*

The nerve-tubes of the ventral (anterior) nerve-roots do not subdivide after they have entered the spinal nerve-cord, according to my observations. I have been able to trace them for considerable distances into the nerve-cord, in chromo-silver stained preparations (vide fig. 105) as well as in carmine and hæmatoxylin preparations (vide fig. 93 and 95) but no subdivisions have been discovered, though the tubes are thick and easy to observe; slender side-branches are, however, given off (vide fig. 93, *nbr*; 105, *sbr*).

In a few cases only have I been able to trace their *direct combination with ganglion cells* (vide fig. 106) and the tubes which in these cases had a direct combination were comparatively very thin.

From what is seen, there can, however, be no objection to the assumption, that all the nerve-tubes of the ventral nerve-roots are directly connected with ganglion cells, indeed, I think this is all but certain, seeing that these tubes have no subdivisions; but it is probable that the thick ones pass for some distance through the white and grey substance before they united with the cells. A great many of them seem to pass over into the other side of the spinal cord crossing the ventral transverse commissure.

I will not omit to mention a section which I once obtained of a chromo-silver stained preparation. A part of this section is illustrated in fig. 102 which is drawn under the camera lucida and is an exact representation of what was seen in the preparation. The nervous process (mpr_1) or perhaps »mixed process« (cf. above) issuing from the ganglion cell *ge*, is seen to give off a thick branch (br_1), which passes towards the ventral (anterior) nerve-root (*vnr*). This branch becomes very thin towards the periphery of the white substance and can not be traced entirely into the ventral nerve-root, though there only wants an extremely short distance. Whether this branch was really connected with the ventral nerve-root (was transformed into a peripheral nerve-tube), and it was only owing to an incomplete staining that it could not be traced into the nerve-root was very difficult to decide. In its whole appearance this branch was not at all like common protoplasmic branches, which I certainly very often have observed to be directed towards the nerve-roots (vide fig. 105, *ppr*; 109, *ppr*; 93, gc_2), but it had that smooth and black staining which is caracteristic for nervous processes. What is strange is, that in the same section (which was somewhat thick) I observed two similar processes (consequently from other cells) running parallel to this one, situated one over the other and giving off quite similar processes at the same place and towards the same nerve-root. I have drawn only one of these processes (mpr_v) in the illustrations in order not to make it too complicated. These 3 processes were thus situated that they quite covered each other vertically; in the illustration I was obliged, of course, to draw them running beside each other. Similar instances of processes running quite the same way, one over the other on the border between the grey and the white substance and giving off branches at certain places are not rare in the spinal cord of *Myxine*.

The longitudinal nerve-tubes of the spinal cord. — I have observed side-branches to be given off from many of the nerve-tubes running longitudinally in the spinal cord (vide fig 111, lnt'; 112, lnt'). Some longitudinal nerve-tubes are also, by sub divisions, broken up into slender fibrillæ (vide fig. 111, snt; 112, snt) whether these tubes are nerve-tubes coming from other parts of the central nervous system, or they are only branches coming from some dorsal nerve-roots, I can not decide. In some instances I saw them cross the transverse commissure and pass over into the other side of the nerve-cord (vide fig. 111, snt').

The neuroglia. — Neuroglia-cells are situated in great abundance in the grey substance of Myxine. Their nuclei are smaller than those of the ganglion cells, have a circular or ovate form, and a somewhat granular appearance. The neuroglia cells have often many processes (fig. 103, nuc_1, nuc_2).[1]) These processes do not subdivide much but, like the protoplasmic processes, the traverse the white substance to its periphery, and terminate there under sheath enveloping the spinal cord (vide fig. 95, f). The course of the processes is particularly easy to study, in successful chromo-silver stained sections where they are seen in great abundance as black, or reddish black, fibres, passing from the grey substance to the periphery, everywhere, through the white substance. The same neuroglia cells sends often, at the same time, processes to the dorsal side as well as to the ventral side of the spinal cord (fig. 103, nuc_1, nuc_2). The processes of the neuroglia cells can, however, easily be traced also in section

---

[1]) It may be that in nuc_1 and nuc_2 several neuroglia-cells, situated close together, have been stained in one body; seeing, however, that the many processes issue from so comparatively very small spaces I think it to be not improbable that we have here single cells (at all events in nuc_1).

stained with carmine or hæmatoxylin[1]) (HEIDENHAIN's hæmatoxylin-staining can especially be recommended). Fig. 101 will probably give a good impression of in what multitudes they occur in the white substance.

The processes of the neuroglia cells do not anastomose, neither those from the same cells, nor those from different cells. Though I have, for instance, seen so many processes issuing close to each other as is illustrated in fig. 108[2]) yet I have not found *a single anastomosis.* In sections, stained, for instance, with HEIDENHAIN's hæmatoxylin (fig. 101), it is seen, as, above mentioned, that there, between the longitudinal nerve-tubes of the white substance, runs a multitude of slender fibres in every direction, but *neither, here, are anastomoses found.* The chief part of these fibres, running between the nerve-tubes, are neuroglia fibres coming from the neuroglia cells and penetrating to the periphery, but some of them are, also, as we have seen, branches of protoplasmic processes coming from the ganglion cells, and finally some fibres are branches from the nervous processes. It is, generally, in sections, stained in the common way (i. e. carmine, hæmatoxylin etc.), exceedingly difficult to draw any distinction between these various kinds of fibres or tubes.

The cells of the epithelium surrounding the central canal resemble, in their appearance, very much, the neuroglia cells, their nuclei have, to some extent, the form and appearance of common neuroglia nuclei. At their exterior extremities, these cells have processes which penetrate, at all events to a great extent, to the periphery of the spinal cord on the dorsal side, as well as on the ventral one (vide fig. 93, 95, 103); these processes are, consequently, quite like those of the neuroglia cells. Some processes from these cells have a more lateral direction into the grey substance (vide fig. 94, *eee*; 102, *pee*) I have not been able to trace these processes to their extremitie.[3])

---

[1]) Vide what is said of that p. 152.

[2]) In this preparation I think it is probable that some cells have been stained at one and same time and have acquired an appearance as if they were one cell. The illustration is drawn to nature as exactly as possible, and it may be seen that there are several irregularities in the staining of the mesial part. Similar preparations are very often obtained.

[3]) In the spinal cord of higher fishes (Teleostii), I have also by chromo-silver staining been able to trace the processes of the epithelial cells of the ventral canal through the white substance and just to its periphery where they terminate or unite with the sheath. As regards the processes of these central epithelial cells in higher Vertebrata I will refer the reader to *Golgi*'s observations (Sulla fina etc., vide list of Literature 1885) he has in the spinal cord of the chicken observed similar processes

The cells situated in the grey substance just outside the epithelial cells, are quite like the latter. They are elongated at their interior extremities, and penetrate between the epithelial cells to the central canal (if they cary cillia, like the latter, I can not decide); at their exterior extremities they have processes wich penetrate to the surface of the spinal cord, like those from the epithelial cells. These cells represent, consequently, a kind of transition-stage between the common neuroglia cells and the epithelial cells surrounding the central canal; indeed, in the many layers of such cells surrounding the epithelium of the canal, we can find all possible transition-stages. When we, now, consider that in *Amphioxus* the neuroglia principally consists of epithelial cells (we do not, here, refer to the sheaths surrounding the nerve-tubes etc.) and there is every reason to believe that this is a primary state from which the conditions found in Myxine are developed, I think we are fully entitled to say, that *the neuroglia cells have an ectodermal origin, and spring from the epithelial cells of the central canal* or we may rather say *the central groove*.[1])

The nerve-cells of the spinal ganglia. — Before we leave the nervous system of *Myxine* I will make some remarks on the structure of the spinal ganglion cells, as I think, they exhibit relations which in several respects are of great interest for the subjects treated of in this paper. Upon the whole the spinal ganglion cells reminds in their structure very much of the ganglion cells of the central nervous system of many invertebrates, and, amongst those we have here examined, especialy the large unipolar cells of *Homarus vulgaris*. They represent obviously a less developed stage than that of the ganglion cells of the spinal cord.

---

traversing the white substance (vide l. c. p. 179). Regarding these processes in higher Vertebrata I vill also refer the reader to the various memoirs by *Ainslie Hollis* (l. c. 1883), *Leydig* (Zelle und Gewebe 1885 p. 189), *Rabl-Rückhard* (l. c. 1883), *Reissner* (l. c. 1864), *John Dean* (l. c. 1863) and others.

[1]) I think it is most probable that the neuroglia cells of the invertebrates have, also, an ectodermal origin. This may, for instance, easily be seen in Nereis, where the ventral nerve-cord is not yet fully separated from the ectoderm. It is, here, even difficult to see any difference between the neuroglia and the ectodermal cells, and it is in many cases impossible to draw any distinct line between the cells enveloping the nerve-cord, and the ectodermal cells situated outside them (vide fig. 14). I have, here, only considered what we may infer from the nervous system of full grown specimens. In examination of the embryological development of the nervous system, the ectodermal origin of the neuroglia will be still more evident; on that subject the reader may be referred to the recent paper by *Kleinenberg* on the embryologie of the Annelides (Zeitschr. f. wiss. Zool. 1886).

Fig. 97, 98 and 99 represent sections through various spinal nerve-cells stained with carmine. *Their sheaths* are in sections rather conspicuous, and there is, generally, several nuclei situated inside (and also outside) them (vide fig. 97, *n*; 98, *n*; 99, *n*). These sheaths and their nuclei resemble very much the sheaths and nuclei of the ganglion cells of Homarus. The nuclei are not, however, present in such number in the latter as the sometimes are in the spinal cells.

The structure of the protoplasm of the spinal nerve-cells is very similar to that of the protoplasm of the common ganglion cells. The lightly staining areas (transsected tubes) are also present in sections of this protoplasm, but they are perhaps still more distinctly prominent and frequently arranged in defined groups. Sometimes they are, in sections, present on two sides of the nuclei in the elongated cells (fig. 97), somitimes they are present only in one side of the cells (fig. 98, 99). On careful examination of successful sections it becomes quite evident that these lightly staining areas are transsected tubes, or perhaps bundles of primitive tubes. These tubes or bundles have generally, when the cells are bipolar or elongated, a longitudinal course through the protoplasm of the cells, and they are often logitudinally transsected (fig. 98, *v*). What these tubes or bundles are is not easy to say, my observations are still too imperfect. In sections of cells where the processes are longitudinally transsected in their origin, it is, however, frequently seen that these tubes or bundles run towards the origin of the processes (vide fig. 98). I think it is probable that they constitute the process in a way similar to what the bundles of primitive tubes do in the ganglion cells of *Homarus* (cf. p. 103). In the bipolar cells it is possible that the contents of the nerve-tube, entering a cell at one end, is broken up into such bundles of primitive tubes which traverse the protoplasm longitudinally and re-unite at the other end to form the peripheral nerve-tube quitting the cell here. As is already known from FREUD's paper (l. c. 1878) there exist unipolar and bipolar ganglion cells in the spinal ganglion of *Petromyzom*, and in this respect as in others they resemble those of *Myxine*.

In the nerve-cells of the spinal ganglia of the frog, M. v. LENHOSSÉK has described »eine seichte, aber breite durch sharfe Ränder begrenzte, tellerförmige Vertiefung, whelche durch 2—3 rundliche, bisher unbekannte Zellen beinahe vollständig ausgefüllt wird« (l. c. 1886 p. 450). This »Vertiefung« was situated in the surface of the cells near the origin of their processes. In some spinal cells of *Myxine* I have observed structures which seem to be very like these.

Fig. 99 represents a cell with a such structure. The light circular area x is very distinctly defined. There is a very long nucleus (N), with some protuberances, situated in one side of this area, in the centre there is another somewhat lighter corpuscle (N') which is perhaps also a nucleus. I could, however, see no cells in this area, and whether it is one cell or contains several cells or what it is, I can not decide.

The nuclei of the spinal nerve-cells are large and exhibit generally a distinct chromatin reticulation.

On comparing the main characters in the structure of the nervous elements of these two vertebrates (*Amphioxus* and *Myxine*) with what is previously described in this paper, we will see that there is the most perfect correspondence in this respect between the invertebrated bilaterates and these inferior vertebrates. But as we can not easily suppose that their is any principal difference between them and higher vertebrata in these respects, we may thus assume that the principal results arrived at in this paper are applicable to all bilaterates. In a future paper I hope to be able to pay a more special attention to the nervous elements of the vertebrata.

6. The combination of the nerve-tubes with each other.

A question of much interest arises, viz. how is the combination, between the nerve-tubes produced, or rather how is the impression received by a sensitive nerve-tube conveyed to a motoric one. To express it in another way me may say, which elements has an irritation of a sensitive nerve-tube to pass, to produce a reflex-movement?

A very common theory is, as is well-known, that an irritation of a sensitive nerve-tube passes to a »sensitive« ganglion cell, then, from this cell, through a protoplasmic process to a motoric ganglion cell, and from that cell it passes into a motoric nerve-tube to produce a reflex-movement. In a previous chapter (ch. 4, p. 148), it is, I hope, suficiently proved, that in the invertebrated bilaterates there is no direct combination between the ganglion cells through the protoplasmic processes, neither is there in *Myxine* (as we have seen in the previous chapter); the above quoted theory of the course of an irritation producing a reflex-movement is therefore for this reason alone not maintainable.

Before we enter into the inquiry of how a combination between the nerve-tubes is produced, we will examine if there is any morphological difference to be observed between motoric nerve-tubes and sensitive nerve-tubes.

As is above stated the nerve-tubes of the dorsal (posterior) nerve-roots of the spinal cord of *Myxine* subdivide, immediately after their entrance into the nerve-cord and thus lose their individuality, whilst the nerve-tubes of the ventral (anterior) nerve-roots retain their individuality and seem to spring directly from ganglion cells, whilst giving off slender side-branches.

I have also observed similar conditions in the spinal cord of higher vertebrata. I have even been able to obtain here more complete observations on the subdivisions of the nerve-tubes of the posterior nerve-roots.

My observations fully agree therefore with those of Prof. C. GOLGI in this respect; and as GOLGI says, I think we are entitled to assume that, in the *spinal nerve-cord, the sensitive nerve-tubes arise from a junction of slender nervous fibrillæ*, whilst the *motoric nerve-tubes are direct continuations of nervous processes* (from ganglion cells) *which however give off slender side-branches*. If this is correct the opposite conclusion seems, also, to be admissible, at all events to a certain extent, viz. that nerve-tubes arising from a junction of nervous fibrillæ are *sensitive nerve-tubes*, whilst nerve-tubes springing directly from ganglion cells are *motoric nerve-tubes*. In the brain of the vertebrata, we find nerve-tubes originating in both ways, and hence, I think, there, is every reason to believe that the same morphological differences between motoric and sensitive nerve-tubes are there present.

Having now settled what morphological difference there appears between the origin of sensitive and motoric nerve-tubes in vertebrata, and seeing that in invertebrata quite a similar difference is present in the modes of origin of two kinds of nerve-tubes, I think we are entitled to conclude further: that in the nervous system of the invertebrated bilaterata there is the same morphological difference between the origin of the sensitive and the motoric nerve-tubes, as is found in vertebrata; viz. the *sensitive* nerve-tubes arise from a junction of nervous fibrillæ, whilst the *motoric* nerve-tubes spring directly from ganglion cells.

Arrived at this conclusion, we will now consider the question, how is an impression, probably, conweyed from a sensitive nerve-tube, to a motoric one? As it is chiefly the invertebrated bilaterata which have been treated of in this paper, we will also in the present inqiury principally confine our remarks to the consideration of their nervous system only.

As is above seen, the sensitive nerve-tube probably subdivides, and is broken up into slender branches, after its entrance into the dotted substance, it is, thus, not very probable, that an irritation of such a nerve-tube should be conweyed to one ganglion cell only, it would then be necessary that all the branches of a sensitive nerve-tube should unite again to form the nervous process of a ganglion cell (sensitive ganglion cell).

According to what is before stated of the combination of the ganglion cells it will be obvious that: to be forwarded to a motoric cell, the irritation would necessarily after its arrival in the sensitive ganglion cell have to return through the same nervous process, to pass over through some other branches of the process into the nervous process of a motoric cell; arrived in the motoric cell it had then to return through the same process and finally pass into the centrifugal nerve-tube, to produce the reflex movement. It needs no further discussion to see that this would be a highly artificial and complicated explanation of the course of an irritation, from a centripetal to a centrifugal nerve-tube. For one thing, it is not, in my opinion, admissible to assume that all the branches arising from the subdivisions of a sensitive tube should re-unite to constitute the nervous process of a sensitive ganglion cell; why then does the subdivision of the nerve-tube exist, why does not the nerve-tube pass undivided to the cell? For another thing, I do not think it admissible that the same irritation should pass and return through the same nervous processes.

In my opinion, it is a much more probable assumption that *the irritation does not at all pass through any ganglion cells,* but that after its arrival in the dotted substance, through subdivisions of the centripetal nerve-tube, it *is at once carried on, into one or several centrifugal nerve-tubes by the slender side-branches or fibrillæ joining them.* How the combination between the branches of the centripetal nerve-tubes and the side-branches of the centrifugal nerve-tubes is produced in the dotted substance, can not at present be defined; I can not, however, see any objection to the assumption that it does exist.

According to the above conclusion, we may thus illustrate a *reflex-curve* (i. e. the way an irritation of a centripetal nerve-tube has to pass to arrive in a centrifugal nerve-tube and produce a reflex-movement) as in the diagram fig. 113, the large arrows indicate the course of an irritation producing a reflex-movement. The *reflex-curve is*, consequently, *composed of the following constituents:* 1) *centripetal* (sensitive) *nerve-tube*; 2) *the central web or interlacing of nervous fibrillæ or tubes* (fig. 113, d); 3) *the centrifugal* (motoric) *nerve-tube*.

This theory will necessarily give a new view of the functions of the central element; but it will, I think, explain a great many relations which have been rather difficult to understand. The dotted substance (the interlacing of nervous fibrillæ) must be a principal seat of the nervous activity, through this substance or interlacing is the reflex-actions etc. communicated to the consciousness, which even possibly has its seat in this substance itself (especially that of the brain). According to this view there can of course, to some extent, be a *localisation* in the central nerve-system but *no isolation*. This view will also I think possibly be able to explain the fact that other parts of the brain can take up the function of lost parts. This is not, however, the place to enter into such physiological details, we have especially taken up the histological side of the question and in this respect we can state, as a fact, that a plaiting or interlacing (not reticulation) of nervous fibrillæ extends through the whole central nervous system of all animals (which posses a central nervous system) and that probably all peripheral nerve-tubes, entering into or issuing from the central nervous system, are connected with this central fibrillar interlacing by branches. We will then ask the physiologists if it is not probable that it is this interlacing of nervous fibrillæ (or tubes) which especially produces the feeling of unity in the nervous system, in other words that it is the principal seat of self-consciousness?

7. The function of the ganglion cells.

A great many physiologists will probably object, that if this is correct, what functions have, then, the ganglion cells?

I have thought of this, and am fully aware that if the proposed theory of the elements of the reflex-curve is approved, it will cause the common theory of the function of the ganglion cells to totter.

The reflex-actions are, as is well known, probably, the primary state or the starting point of a great many, if not all, actions (voluntary or unvoluntary.)

And if reflex movements are produced without direct help of the ganglion cells, then there are certainly a great many actions which can be produced without any direct assistance from these cells.

In the same way as above, me may also assume that an impression or irritation is conweyed from one part of the central nervous system to another, e. g. from the ventral (or spinal) nerve-cord to the brain, or vice versa, without any passage through ganglion cells; an irritation producing a voluntary action may, thus, be conweyed from the brain and into the centrifugal nerve-tubes, passing to the active limb, without passing any ganglion cells on the way, or without any direct action of the ganglion cells of the ventral (or spinal) nerve-cord, or of the motoric centre of the limb. It is not in our present state of knowledge possible to predict the extent of the influence this theory, if approved, will have an our ideas about the function of the nervous system. I will, therefore, by no means attempt to trace it here, but will subsepuently return to the ganglion cells, and examine what functions there may be left for them, for certainly we cannot one moment doubt that they are of great importance to the nervous system.

A function which I suppose they have, and which certainly is a very important one, is to serve as nutritive centres of this nerve-tubes and fibrillæ originating in them. We have, before, seen, that the protoplasmic processes have probably a nutritive function and absorb nutrition, for the cell, in the periphery of the ganglia (or nerve-cords) or near the walls of the blood-vessels; now, I think it is probable that this nutrition is not meant to be for the cell only, but that it is assimilated by the protoplasm of the cell (the nucleus has probably an important function in this respect) and is, in a suitable form, carried on into the primitive tubes of the nervous processes; in other words, the ganglion cells are the nutritive centres of the nervous processes, and then probably also of the branches and nerve-tubes arising from these processes.[1]) We know of what great importance nutrition is for the nervous system; it is very peculiar, in

---

[1]) Those ganglion cells, from which the motoric nerve-tubes directly issue, are the nutritive centres of these tubes; whilst those ganglion cells, the nervous process of which are entirely broken up into slender branches, are the nutritive centres of these branches (which contribute to the formation of the dotted substance).

this respect, and it is well known that a few moments want of nutrition for the brain is enough to disorder the nerve-system and produce unconsciousness in man. The nutrition is certainly not a function of too insufficient importance to explain the extension the ganglion cells have in the central nervous system.

That the ganglion cells of the spinal ganglia have a nutritive function is in my opinion vere probable. As the nerve-tubes of the dorsal (posterior) nerve-roots have got no special nutritive centres in the central nerve-system (as they do not originate directly from cells) they have got them in special ganglia outside the central nerve-system. In *Myxine* the spinal nerve-cells are unipolar or bipolar. *The processes of the unipolar cells subdivide dichotomically in a T* and send one branch peripherically and one centrally (cf. FREUD's investigations l. c. 1878). In my opinion the cell is the nutritive centre of both these branches, but is of no direct importance for their sensitive function; the irritation of the peripheral branch can not pass through the cell (as it then must pass and return through the same process) but passes directly into the centripetal branch at the subdivision in T of the process. In the *bipolar cells* the irritation of the peripheral part of the nerve-tube must certainly pass through the cell to come into the centripetal part; in so far it is no objection that the bipolar cell may be of direct importance for the sesitive function of the nerve-tubes. But the fact only that in the same ganglion, *unipolar* cells are present and are united with quite similar nerve-tubes must be enough to make us doubtful in this respect. If the unipolar cells have a nutritive function only, it must be probable that the bipolar cells have the same function; indeed, seeing that in higher verterbrata the spinal ganglia consist exclusively of unipolar cells (notwithstanding the slender protoplasmic processes described by FRITSCH (l. c. 1886) which processes I suppose to be nutritive processes, cf. above p. 146, note 1), I think that this must be quite certain. There are also appearances in the structure of the bipolar cells, which, as we have before seen, indicates that the contents of the processes to some extent pass through the protoplasm of the cells without intermixing with it (compare the description of the protoplasm of these cells p. 163).

We arrive thus at the conclusion that *all the nerve-cells of the spinal ganglia have a nutritive function only*. They are probaly of a similar importance for the sensitive (centripetal) nerve-tubes as the »motoric« ganglion cells (i. e. the cells from which the motoric nerve-tubes spring) are for the motoric (centrifugal) nerve-tubes, the difference is that the former are situated outside the central nervous system and have got a simple shape without many prosesses, probably because they can easily get their nutrition without special organs; whilst the latter are situated in the central nervous system and need generally many and long processes to reach the bloodvessels or the periphery of the nervous organs and there obtain sufficient nutrition for the cell and its nervous process.

From the recent works by my much estimated friend ADOLF D. ÓNODI (l. c. 1884, 1885, 1886) (cf. also the previous papers by MARSHALL, l. c. 1877; BALFOUR, l. c. 1876 & 1877; HENSEN, l. c. 1876; SCHENK, l. c. 1876) we have learnt that the spinal ganglia spring from the spinal cord. We are thus entitled to say there is no principal difference between the *ganglion cells* of the nervous system and those of the spinal ganglia, *they have the same origin and the same function* but they have got a different situation and *differ therefore also in shape, some are unipolar whilst other are bipolar, and even multipolar, which, however, does not make any difference in their function.*

I have not yet had opportunity to examine in invertebrates how the relations are as to the nutritive centres of those nerve-tubes which spring from the dotted substance by a union of slender tubes. Whether the ganglion cells which I have found in the peripheral nerves of various animals (cf. my paper on Mysostoma 1885 p. 31 & 74 and this paper p. 141 & 143, note 2) are of importance in this respect I can not decide at present, it is, at all events, not impossible. In the peripheral nerves of Patella I have, as mentioned p. 141 found ganglion cells which had even protoplasmic processes (cf. fig. 20, *gc*).

Whether the ganglion cells of the central nerve-system have any functions besides being nutritive centres is, of course, extremely difficult to decide.

It is not imposisble that they may be the seat of *memory*. A small part of each irritation producing a reflex action, may on its way through the dotted substance be absorbed by some branches of the nervous processes of the ganglion cells, and can possibly in one way or another be stored up in the latter. Howsoever that may be, and whatever the function of the ganglion cells is, this new theory of the combination between the centripetal and the centrifugal nerve-tubes gives, if approved, a quite new view of the importance of the dotted substance (or the interlacing of nervous fibrillæ of the vertebrates) and will, in my opinion, explain many facts as to its occurrence. If the theory is correct, then, the dotted substance must be a principal seat of the nervous activity, and — the higher an animal is mentally developed — the more complicated and extensive must we expect to find its dotted substance; this is in the fullest harmony with the facts already known. We need only refer the reader to the ant, or the bee, to compare the extremely complicated and highly developed dotted substance of these small inteligent animals, with the dotted substance of less developed insects, or to compare the dotted substance of the insects or crustacea with that of annelides, etc. etc. I am sure that my readers will very soon arrive at the conclusion, that the more complicated the structure of dotted substance is — the more highly is the animal mentally developed; in other words, we may conclude that *the more the inteligence of an animal is developed — the more intricate becomes the web or plaiting of nerve-tubes and fibrillæ in its dotted substance;* the protoplasmic processes etc. of the ganglion cells are of no great importance in this respect. In this manner we can explain how it is, that unipolar cells occur in the nervous system of animals (e. g. insects and crustacea) which are mentally even highly developed; these animals have an extremely intricate web of nerve-tubes in their dotted substance, and this web is probably the principal seat of inteligence.

List of the principal Literature

having reference to the subjects treated of in this memoir.

(There is also referred to a few papers to which I have had no access as we have not got them in the library of Bergen's Museum, and which I only know from the reports by other authors.)

1804. Mangili. Nuove richerche zootomiche sopra alcune specie di Conchiglie bivalvi. Milano.

1807. A. Barba. Osservazioni microscopiche sul cervello e sue adjacenti. Napoli. An abstract in Reil's Archiv. Bd. X. 1811 p. 467.

1819. — Osservazioni microscopiche etc. . . Seconda edizione aggiunte. Napoli.
— Translated in German: (Beobachtungen über das Gehirn etc.) by Dr. J. J. Albrecht v. Schönberg. Würzburg 1829.

1820. Treviranus. Vermischte Schriften anat. und phys. Inhalts. Bd. III. Untersuchungen über den Bau und die Funktionen des Gehirns der Nerven und der Sinneswerkzeuge in den verschiedenen Klassen und Familien des Thierreiches. Göttingen.

1824. Serres. Anatomie comparée du cerveau.

1825. Roth. De animalium invertebratorum systemate nervoso. Disertatio inaugural Wiceburgi.

de Blainville. Manuel de Malacologie.

1828. Milne Edwards. Recherches anatomiques sur le système nerveux des Crustacés. Ann. d. sci. nat. T. XIII p. 113—123.

Johannes Müller. Ueber ein eigenthümliches, dem Nervus sympathicus analoges Nervensystem der Eingeweide bei dem Insekten. Nova acta Leop. Carol. nat. cur. T. 14. Pars I. Bonn p. 71—108.

1829. Morren. Histoire du Lombric terrestre. Bruxelles.

Lyonet. Anatomie de differentes espèces d'Insectes. Mem. du museum. Paris Tom. 18.

1832. George Newport. On the nervous system of the Sphinx ligustri, Lin. and on the changes which it undergoes during a part of the metamorphoses of the insect. Phil. trans. of the Royal Soc. London 1832 p. 383.

1833. Christian Gottfried Ehrenberg. Ueber die Structur des Gehirns und der Nerven. Froriep's Notizen. XXXVII.
— Nothwendigkeit einer feineren mechanischen Zerlegung des Gehirns und der Nerven vor der chemischen, dargestellt aus Beobachtungen. Poggendorfs Annalen der Phys. Bd. XXVIII p. 449—472.

1834. — Bemerkungen zum vorhergehenden Aufsatz. (Bemerkungen über die feinsten Nervenfasern, von Krause.) ibid.
— Beobachtungen einer bisher unbekannten auffallenden Structur des Seelenorgans bei Menschen und Thieren. Abh. der kgl. Akad. Wiss. Berlin 1834 p. 665—722.

G. G. Valentin. Ueber die Dicke der varikösen Fäden in dem Gehirn und dem Rückenmark des Menschen. Müller's Arch. f. Anat. u. Phys. 1834 p. 400—409.

Milne Edwards. Histoire naturelle des Crustacés. T. I. p. 120. Paris.

George Newport. On the nervous system etc. (Part II.) during the latter stages of its pupa and its imago state; and on the means by which its development is effected. Phil. trans. R. Soc. London 1834 p. 389.

Krohn. Über die Verdauungsnerven des Flusskrebses. Isis 1834 p. 530.

1835. Berres. Mikroscopische Beobachtungen über die innere Bauart der Nerven und Centraltheile des Nervensystems. Midicinischen Jahrbüchern d. k. k. österreichischen Staats 1835 p. 274.

1836. Langenbeck. De retina observationes anatomicae-pathologicae. Gottingae.

Brandt. Remarques sur les nerfs stomato-gastriques ou intestinaux dans les animaux invertébrés. Ann. des sci. nat. Ser. 2 T. V. p. 81—110. Translated from the German memoir in: Memoires de l'Academie Imperiale des Science de Saint-Petersbourg. Tom. III.

Gabr. Gust. Valentin. Ueber den Verlauf und die letzten Enden der Nerven. Nova acta Leop. Carol. nat. cur. Tom. XVIII p. 51—240, 541—543.

G. R. Treviranus. Beiträge zur Aufklärung der Erscheinungen und Gesetze des organischen Lebens. 1836.

Robert Remak. Vorläufige Mittheilungen microscopischer Beobachtungen über den inneren Bau der Cerebrospinalnerven und über die Entwickelung ihrer Formelemente. Müller's Arch. f. Anat. u. Phys. 1836 p. 145—161.

1837. — Weitere mikroskopische Beobachtungen über die Primitivfasern. Froriep's neue Notizen. Bd. III p. 39 (vide also ibid p. 137).

Garner. On the nervous system of molluscous animals. Trans. of the Linn. Society, London Vol. XVII p. 485.

Johannes Evangelista Purkinje. Untersuchungen über Nerven und Hirnanatomie. Ber. ü. d. Versamml. d. deutsch. Naturf. u. Aerzte (Prag) 1837 p. 177—178.

— Ueber die gangliösen Körperchen in verschiedenen Theilen des Gehirns. ibid. p. 179.

1838. Johannes Müller. Vergleichende Anatomie der Myxinoiden. Neurologie. Abhandl. d. königl. Akad. d. Wiss. zu Berlin 1838.

Robert Remak. Observationes anatomicae et microsc. de systematis nervosi structura. Diss Berolini.

Volkmann. Ueber die Faserung des Rückenmarkes und des sympatischen Nerven in Rana esculenta. Müller's Arch. f. Anat. u. Phys. 1838.

1839. Carpenter. Dissertation on the physiological inference, to be deduced from the structure of the nervous system in the invertebrated classes of animals. Edinburgh.

Georg Newport. Insects. Todds Cyclopædia of Anatomy and Physiology. Vol. II p. 942—960.

Anderson. Nervous system. Todds Cyclopædia of Anatomy and Physiology vol. III. London.

Gabr. Gust. Valentin. Ueber die Scheiden der Ganglienkugeln und deren Fortsetzungen. Müller's Arch. f. Anat. u. Phys. 1839 p. 139—165.

— Zuv Entwickelung der Gewebe des Muskel-, Blut- und Nervensystems ibid. p. 194.

— De functionibus nervorum cerebralium et nervi sympathici. Bernæ Liv. I.

Rosenthal. De formatione granulosa in nervis etc. Vratislaviae.

1841. Robert Garner. On the anatomy of the Lamellibranchiate Conchifera. Trans. of the Zool. Soc. of London. Vol II p. 87.

Grant Outlines of Comparative Anatomy p. 185—204.

1842. Milne Edwards. Sur la structure et les fonctions de quelques zoophytes. mollusques et crustacés des côtes de la France. Ann. des sci. nat. Ser. 2 T. XVIII. Nerv.syst. p. 326—329.

Helmholtz. De fabrica systematis nervosi evertebratorum. Diss. inaug. Berolini (vide also reference in Siebold & Reichert. Arch. f. Anat. u. Phys. 1843 p. 11 & CXCVII).

Hannover. Mikroskopiske Undersøgelser af Nervesystemet. Kjøbenhavn.

1843. A. de Quatrefages. Memoire sur l'Éolidine paradoxale. Ann. d. sc. nat. Ser. 2. T. XIX p. 293—300.

Robert Remak.[1]) Über die Inhalt der Nervenprimitivröhren. Müller's Arch. f. Anat. u. Phys. 1843 p. 197—201.

1844. — Neurologische Erlänterungen. Müller's Arch. f. Anat. u. Phys. 1844 p. 463—473.

Hannover. Recherches microscopiques sur le système nerveux. Copenhague, Paris, Leipzig.

Friedrich Will. Vorläufige Mittheilungen über die Structur der Ganglien, und den Unsprung der Nerven bei Wirbellosen Thieren. Müller's Arch. f. Anat. u. Phys. 1844 p. 76—94.

---

[1]) Vide also Remak. Carper's med. Wochenschrift 1839. Ammon's Zeitschr. f. Med. etc. Bd. III. Hft. 3. 1840. Med. Vereins-Zeitung. 1840 No. 2.

1844. A. W. Volkmann. Ueber Nervenfasern und deren Messung mit Hülfe der Schrauben- und Clasmikrometer. Müller's Arch. f. Anat. u. Phys. 1844 p. 9—26.

Julius Budge. Ueber den Verlauf der Nervenfasern im Rückenmarke des Frosches, ibid. 1844 p. 160—190.

G. Valentin. Erwiderung auf . . . Volkmann'schen Aufsatz über Nervenfasern etc. ibid 1844 p. 395—404.

F. Bidder und A. W. Volkmann. Die Selbständigkeit des sympathischen Nervensystems durch anatomische Untersuchungen nachgewiesen.

F. Bidder. Erfahrungen über die functionelle Selbständigkeit des sympathischen Nervensystems, aus brieflichen Mittheilungen von F. Bidder an A. W. Volkmann. Müller's Arch. f. Anat. u. Phys. 1844 p. 358—381.

1845. A. de Quatrefage. Sur le système nerveux et sur l'histologie du Brachiostome ou Amphioxus. Ann. d. sci. nat. Ser. 3. T. IV p. 214—229.

Emile Blanchard. Recherches anatomiques et géologiques sur l'organisation des Insectes et particulièrement sur leur système nerveux. Premier partie: les Coléoptères. Comptes Rendus. Paris. XXI p. 752—754, 963—964.

— Observations sur le système nerveux des Mollusques testacés ou Lamellibranches. Ann. d. sc. nat. Ser. 3. T. III p. 321—340.
(vide also Froriep Notizen XXXIV 1845. Comptes rendus. Paris XX 1845 p. 496—489.)

— Recherches sur le système nerveux des Mollusques gastéropodes. Extr. procès-verbaux de la Soc. philom. Paris. 1845 p. 25—27.

1846. — Du système nerveux des insectes. Mémoire sur les Coléoptéres. Ann. d. sc. nat. Ser. 3. T. V p. 273—379.

Harless. Briefl. Mittheilung über die Ganglienkugeln von Torpedo Galvanii. Müller's Arch. f. Anat. u. Phys. 1846 p. 283.

H. Lebert und Ch. Robin. Kurze Notiz über allgemeine vergleichende Anatomie niederer Thiere. Müller's Arch. f. Anat. u. Phys. 1846 p. 128—129.

1847. Rudolph Wagner. Ueber den feinen Bau des elektrischen Organes im Zitterrochen.

— Neue Untersuchungen über die Elemente der Nervensubstanz.

Bidder. Zur Lehre von dem Verhältnisse der Ganglienkörper zu den Nervenfasern. Leipzig.

Axmann. De gangliorum structura penitiori eiusque functionibus.

Ch. Robin. Sur la structure des ganglions nerveux des Vertébrés. Extr. procès-verbaux de la Soc. philomatique. Paris. 1847, p. 23—30.

— Sur la structure des ganglions nerveux des Raies. ibid. p. 68—71; Froriep Notizen II & III. 1847.

1848. Emile Blanchard. Du système nerveux chez les Invertebrés (Mollusque et Annelés) dans ses rapportes avec la classification de ces animaux. Comptes rendus. Paris. XXVII p. 623—625.

1849. Lieberkühn. De structura gangliorum penitiori. Berolini.
  Franz Leydig. Zur Anatomie von Piscicola geometrica mit theilweiser Vergleichung anderer einheimischer Hirudineen. Zeits. f. wiss. Zool. Bd. I p. 129—131.
  A. Kölliker. Neurologische Bemerkungen. Zeits. f. wiss. Zool. Bd. I p. 135—163.
  Carl Bruch. Ueber das Nervensystem des Blutegels. Zeits. f. wiss. Zool. Bd. I. p. 164—175.
  Hermann Stannius. Das peripherische Nervensystem der Fische. Rostock.
1850. Rudolph Wagner. Neurologische Untersuchungen. Nachrichten v. d. Georg-Augusts Universität u. d. K. Gesellsch. d. Wiss. Göttingen. 1850 p. 41—56; 1851 p. 185—196; 1853 p. 59—72. Abstract in: Ann. d. sci. nat. Ser. 3. T. XIX p. 370—379. 1853.
  Johann Czermák. Verästelungen der Primitivfasern des Nervus acusticus. Zeits. f. wiss. Zool. Bd. II p. 105—109.
  Felix Dujardin. Memoire sur le susteme nerveux des Insectes. Ann. d. sci. nat. Ser. 3. T. XIV p. 195—206.
  A. de Quatrefages. Memoire sur le système nerveux des Annelides. Ann. d. sc. nat. Ser. 3. T. XIV p. 329—398.
  Clarke. Researches into the structure of the spinal cord. Phil. trans. of the Royal. Soc. of London for 1851 (read Decbr. 5 1850).
1851. Franz Leydig. Ueber Artemia salina und Branchipus stagnalis. Zeits. f. wiss. Zool. Bd. III p. 290—295.
  — Anatomische Bemerkungen über Carinaria, Firda und Amphicora, ibid. Bd. III p. 325—332.
1852. — Anatomisches und Histologisches über die Larve von Corethra plumicornis, ibid. Bd. III p. 438—442.
  — Zur Anatomie und Entwickelungsgeschichte der Lacinularia socialis, ibid. Bd. III p. 457—460.
  A. de Quatrefages.[1]) Mémoires sur le système nerveux, les affinités et les analogies des Lombrics et des Sangsues. Ann. d. sc. nat. Ser. 3. T. XVIII p. 167—179.
  Léon Dufour. Aperçu anatomiques sur les Insectes lepidoptères. Compte rendus. Paris 1852 p. 749.
  Johann Marcusen. Zur Histologie des Nervensystems. Bull. acad. imp. d. sci. St. Petersb. X. p. 187—192.
  G. Schilling. De medullae spinalis textura, Diis. inaug. Dorpat. (vide Müller's Arch. f. Anat. u. Phys. 1853 p. 66).
1853. Heinrich Müller. Bau der Cephalopoden. Zeits. f. wiss. Zool. Bd. IV p. 344.
  Axmann. Beiträge zur mikroskopischen Anatomie und Physiologie des Ganglien-Nervensystems. Berlin.

---

[1]) Vide also Quatrefages: Règne animal. Annelides Atlas Anatomie Pl. 1 d & 1 c.

1853. Franz Leydig. Zur Anatomie von Coccus hesperidum. Zeits. f. wiss. Zool. Bd. V 1853 p. 5—8.
1854. — Ueber den Bau und die systematische Stellung der Räderthiere, ibid. Bd. VI p. 83—87.

R. Remak. Ueber multipolare Ganglienzellen. Ber. ü. Verhandl. d. k. preuss. Akad. Berlin 1854 p. 29.

R. Wagner. Neurologische Untersuchungen.

P. Owsjannikow. Disquisitiones microscopicæ de medullae spinalis textura imprimis in Piscibus factitatæ. Dorpat.

Georg Meissner. Beiträge zur Anatomie und Physiologie von Mermis albicans. Zeits. f. wiss. Zool. Bd. V p. 220—236.

1855. — Beiträge zur Anatomie und Physiologie der Gordiaceen. Zeits. f. wiss. Zool. Bd. VII p. 20—28 & 93—103.

Leconte et Faivre. Etudes sur la constitution chemique du système nerveux de la Sangsue. Gazette medical No 45 1855 p. 709.

Wedl. Untersuchungen über das Nervensystem der Nematoden. Ber. Akad. Wiss. Wien. Bd. XVII.

Franz Leydig. Zum feineren Bau der Arthropoden. Arch. f. Anat. u. Phys. 1855 p. 398.

Metzler. De medullae spinalis avium textura. Diss. inaugur. Dorpat.

Jakubowitsch und Owsjannikow. Mikroskopische Untersuchungen über die Nervenursprunge im Gehirn. Mélanges biologiques T. II. St. Petersburg 1858 p. 333—335. (Bull. phisico—math. d. l'Acad. imp. d. sci. St. Petersb. T. XIV No. 12.)

E. Faivre. Observations histologiques sur le grand sympathique de la Sangsue medicinale. Ann. d. sc. nat. Ser. 4, T. IV p. 249—261.

1856. — Etudes sur l'histologie comparée du système nerveux chez quelques Annélides. Ann. d. sc. nat. Ser. 4, T. V p. 337—374; T. VI 1856 p. 16—82.

Jakubowitsch. Microskopische Untersuchungen über die Nervenursprunge im Rückenmarke und verlängertem Marke, über die Empfindungszellen und sympatischen Zellen in denselben und über die Structur der Primitivnervenzellen, Nervenfasern und der Nerven überhaupt. Melanges biologiques. Bulletin de l'academie imp. des Sciences de St. Petersbourg Tom II p. 374—387.

Benedict Stilling. Ueber den feinern Bau der Nervenprimitivfaser und der Nervenzelle. Frankfurt a. M.

Ernst Häckel. De telis quibusdam astaci fluviatilis. Diss. Berolini 1856.

1857. — Ueber das Gewebe des Flusskrebses. Arch. f. Anat. u. Phys. 1857 p. 469—486 & 532—541.

Lockhart-Clarke. On the nervous system of Lumbricus terrestris. Proc. of the Royal Soc. London. Vol. VII 1856—1857 p. 343.

Jakubowitch. Mittheilungen über den feinern Bau des Gehirns und Rückenmarks. Breslau 1857.

1857. Franz Leydig. Lehrbuch der Histologie des Menschen und der Thiere. Frankfurt a. M.
1858. Jakubowitsch. Recherches comparatives sur le système nerveux. Comptes rendus. Paris. XLXII. p. 290—294, 380—383 (vide also Berliner »Medic. Centralzeitung« 1858 3 Novbr. 88 Stück).

G. Wagener. Ueber den Zusammenhang der Kernes und Kernkörpers der Ganglienzellen mit dem Nervenfaden. Zeitschr. f. wiss. Zool. Bd. VIII 1857 p. 455—457.

J. Gerlach. Mikroskopische Studien aus dem Gebiete der menschlichen Morphologie. Erlangen.

Emile Blanchard. Du grand sympatiques chez les animaux articulés. Ann. d. sci. nat. Ser. 4, T. X p. 5—10. (Comptes rendus XLVII p. 992—995 1858.)

— L'organisation du règne animal. Classe des acéphales, nerv. syst. p. 23. Classe des Arachnides, nerv. syst. p. 39, 110, 149, 186.

1859. Max Schultze. Observationes de retinae structura penitiori. Bonn.

Benedict Stilling. Neue Untersuchungen ueber den Bau des Rückenmarks. Cassel 1859 p. 701—775.

A. Kölliker. Handbuch der Gewebelehre. 3 Aufl.

Jakubowitsch. Études sur la structures intime du cerveau et de la moelle épinière. Ann. d. sci. nat. T. XII. p. 189—245.

1860. Philipp Owsjannikow. Recherches microscopiques sur les lobes olfactifs des Mammifères. Comptes rendus. Paris. L. 1860, p. 428—434. Reichert's Arch. 1860 p. 469—477.

Reissner. Beiträge zur Kenntniss vom Bau des Rückenmarkes von Petromyzon fluviatilis. Arch. f. Anat. u. Phys. 1860.

Henry S. Wilson. The nervous system of the Asteridae. Transact. Linnean Society. London. Vol. XXIII Part the first p. 107. 1860.

Mauthner. Beiträge zur nähern Kenntniss der morphologischen Elemente des Nervensystems. Sitzb. d. kais. Akad. Wiss. Wien. 1860.

John Dean. Microscopic anatomy of the lumbar enlargement of the spinal cord. American Academy of Arts and Sciences 1860.

Franz Leydig. Naturgeschichte der Daphniden.

Owsjannikow. Ueber die feinere Structur der Lobi olfact. der Säugethiere. Arch. f. Anat. u. Phys. 1860.

1861. — Sur la structure intime du système nerveux du Homard. Comptes rendus. LII 1861 p. 378—381.

Stieda. Rückenmark und einzelne Theile des Gehirns von Esox lucius. Dorpat.

Owsjannikow. Recherches sur la structure intime du système nerveux des Crustacés et principalement du Homard. Ann. d. sc. nat. Ser. 4, T. XV p. 128—141.

M. Schultze. Die kolbenförmigen Gebilde in der Haut von Petromyzon etc. Arch. f. Anat. u. Phys. 1861 p. 285.

1862. M. Schultze. Bau der Geruchsschleimhaut. Aus dem 7ten Bande der Abhandlungen der naturforsch. Ges. zu Halle. 1862 p. 65 u. 66 Ann.

J. L. Clarke. Ueber den Bau des Bulbus olfactorius und der Geruchsschleimhaut. Zeitschr. f. wiss. Zool. Bd. XI p. 31—42.

Franz Leydig. Über das Nervensystem der Anneliden. Arch. f. Anat. u. Phys. 1862 p. 90.

1863. R. Buchholz. Bemerkungen über den histologischen Bau des Centralnervensystems der Süsswassermollusken. Arch. f. Anat. u. Phys. 1863.

Walter. Mikroscopische Studien über das Centralnervensystem wirbelloser Thiere. Bonn.

Waldeyer. Untersuchungen über den Ursprung und den Verlauf des Achsencylinders bei Wirbellosen und Wirbelthieren. Henle und Pfeufer's Zeitschr. für ration. Medizin. 3 Reihe. Bd. XX, p. 193—256.

ames Rorie. On the Anatomy of the nervous system in Lumbricus terrestris. Quart. Journ. of Micr. Sci. Ser. 2. Vol. III, p. 106—109.

Ph. Owsjannikow. Ueber die feinere Structur des Kopfganglions bei den Krebsen, besonders beim Palinurus locusta. Mem. Acad. imp. d. sci. St. Petersbourg. VI. No. 10. 1863.

— Ueber die Inauguraldissertation des Herrn Dr. Kutschin das Rückenmark der Neunauge betreffend, nebst einigen eigenen Beobachtungen über das Rückenmark der Knochenfische und anderer Thiere. Novemb. 1863. Mélanges biologiques T. IV St. Petersb. 1865 p. 527—538. (Bull. Acad. imp. sci. St. Petersb. T. VII p. 137—145.)

— Ueber die feine Structur des Kleinhirns der Fische. Dec. 1863. Mélang. biolog. T. IV. p. 551—562. (Bull. Acad. imp. sc. St. Petersb. T. VII p. 157—166.)

A. Kölliker. Handbuch der Gewebelehre des Menschen. 4 Aufl. Leipzig. (5 Aufl. 1867.)

John Dean. The gray substance of the medulla oblongata and trapezium. Smithsonian contributions to knowledge. 1863.

Lionel Smith Beale. On the structure and formation of the so-called apolar, unipolar, and bipolar nerve-cells of the Frog. Philos. trans. Royal. Soc. London. 1863. Vol. 153, p. 543—571. (Abstract in Quart. Journ. micr. sci. Ser. 2. Vol. III p. 302—307.)

1864. Johann Marcusen. Sur l'anatomie et l'histologie du Branchiostoma lubricum, Costa (Amphioxus lanceolatus, Yarrell). Comptes rendus de l'Acad. d. sc. de Paris. LVIII p. 479—483; LIX 1864 p. 89—90. (Ann. Mag. Nat. Hist. XIV 1864 p. 151—154, 319—320.)

Benedict Stilling. Untersuchungen über den Bau des kleinen Gehirns des Menschen. Heft. 1. Cassel.

Schramm. Neue Untersuchungen über den Bau der Spinalganglien Würzburg.

1864. Franz Leydig. Vom Bau des thierischen Körpers. Bd. I. Tübingen.
— Tafeln zur vergleichenden Anatomie. Tübingen.
S. Trinchese. Recherches sur la structure du système nerveux des Moll. gastéropodes pulmonés. Rapport. p. Milne Edward et Blanchard. Compt. rend. T. LVIII 1864 p. 355—358.
Fromann. Ueber die Färbung der Axen- und Nervensubstanz der Rückenmarks durch Argent. nitric. und über die Structur der Nervencellen. Virchow's Arch. XXXI 1864.
Reissner. Über den Bau des centralen Nervensystems der ungeschwänzten Batrachier. Dorpat 1864.

1865. — Zur Structur der Ganglienzellen der Vorderhörner. Virchow's Arch. XXXII. 1865.
Arnold. Ueber die feineren histologischen Verhältnisse der Ganglienzellen des Sympathicus des Froschs. Virchow's Archiw Bd. XXXII. 1865.
Franz Leydig. Zur Anatomie und Physiologie der Lungenschnecken. Arch. f. mikr. Anat. Bd. I. Nerv. syst. p. 44—52.
A. de Quatrefages. Histoire Naturelle des Annelés marins et d'eau douce. Annélides et Gephyriens. Paris. Nerv. syst. p. 77—88.
Lionel S. Beale. Indications of the patles taken by the nerve-currents, as they traverse the caudate nerve-cells of the spinal cord and encephalon. Quart. journ. micr. sci. Ser. 2, Vol. V. p. 90—96.
Ray Lancaster. On the anatomy of the Eartworm. (Quart. journ. of micr. sci. Ser. 2. Vol. V. p. 110—114.)
Baudelot. Observations sur la structure du système nerveux de la Clepsine. Ann. des. sci. nat. Ser. 5. T. III p. 127—136.
Otto Deiters. Untersuchungen über Gehirn und Rückenmark des Menschen und der Säugethiere. Herausgegeben von Max Schultze. Braunschweig.
Luys. Recherches sur le système nerveux cerebro-spinale. Paris.

1866. Jules Cheron. Recherches pour servir a l'histoire du système nerveux des Cephalopodes dibranchiaux. Ann. d. sci. nat. Ser. 5. T. V. 1866 p. 1—118.
Ph. Owsjannikow und A. Kowalevsky. Ueber das Centralnervensystem und das Gehörorgan der Cephalopoden. Mem. d. Acad. Sci. St. Petersb. XI. 1868 (No. 3).
Kutschin. In Referate aus der russischen Literatur von Stieda. Arch. f. mikr. Anat. Bd. II.
Guye. Die Ganglienzellen des Sympathicus beim Kaninchen. Med. Centralblatt. Ser. V. 56 p. 881.
Kollmann und Arnstein. Die Ganglienzellen des Sympathicus. Zeitschr. f. Biologie. Bd. II. Heft. II p. 271.
Polaillon. Etudes sur la texture des ganglions nerveux periphérique. Journ. de l'anat. et de la phys. T. III. 1866.

1867. Ph. Owsjannikow. Ueber das Centralnervensystem des Amphioxus lanceolatus. Mélanges biologiques. T. VI. St. Petersb. 1868 p. 427—450.

Fritz Ratzel. Beiträge zur Anatomie von Enchytraeus vermicularis, Henle. I. Eigenthümliches Schlundnervensystem. Zeitschr. f. wiss. Zool. Bd. XVIII p. 99—102.

Lockhart Clarke. On the Structure of the optic lobes of the Cuttlefisch. Phil. trans. Royal Soc. London. Vol. 157, Part I, p. 155—159.

Ludwig Stieda. Studien über das centrale Nervensystem der Knochenfische. Zeitsch. f. wiss. Zool. Bd. XVIII. 1867 p. 1—70.

— Studien über das centrale Nervensystem der Vögel und Säugethiere. Zeits. wiss. Zool. Bd. XIX. 1867 p. 1—70.

Friedr. Jolly. Ueber die Ganglienzellen des Rückenmarks. Diss. inaug. Zeits. f. wiss. Zool. Bd. XVII p. 443—460.

Fräntzel. Beiträge zur Kenntniss von der Structur der spinalen und sympathischen Ganglienzellen. Virchows Archiv. Bd. 38.

1868. Ernst Ehlers. Die Borstenwürmer. Leipzig 1864—1868.

Victor Lemoine. Recherches pour servir a l'histoire des systèmes nerveux musculaire et glandulaire de l'écrevisse. Ann. de sci. nat. Ser. 5. T. IX. Nerv. syst. p. 100—222.

Schwalbe. Ueber den Bau der Spinalganglien nebst Bemerkungen über die sympathischen Ganglienzellen. Arch. f. mikr. Anat. Bd. IV. p. 45—72.

Courvoisier. Ueber die Zellen der Spinalganglien, sowie des Sympathicus beim Frosch. Arch. f. mikr. Anat. Bd. IV.

Eduard Brandt. Ueber das Nervensystem der gemeinen Schüsselschnecke (Patella vulgaris) and

— über das Nervensystem von Chiton (Acanthochites) fascicularis. Mélanges biologiques. Tom. VII p. 27—40. St. Petersbourg Nov. 1868. (Bull. de l'acad. d. sc. de St. Pet. T. XIII p. 457—466).

Max Schultze. Observationes de structura cellularum fibrarumque nervearum. Bonner Universitätsprogram 1868.

Grandry. Recherches sur la structure intime du cylindre de l'axe et des cellules nerveuses. Bull. de l'academie royale du Belgique. Mars 1868.

Trinchese. Memoria sulla Struttura del Sistemo Nervoso dei Cefalopodi. Firenze.

1869. Edouard Claparède. Histologische Untersuchungen über d. Regenwurm. Zeitschr. f. wiss. Zool. Bd. XIX p. 585—599.

Boll. Beiträge zur vergleichenden Histologie des Moluskentypus. Arch. f. mikr. Anat. Suppliment. Bd. V. Nerv. syst. p. 19.

1870. Fleischl. Ueber die Wirkung von Borsäure auf frische Ganglienzellen. Sitzungsber. d. k. Akad. d. Wiss. Wien. Bd. LXI, 2 Abth., p. 813—818.

C. Golgi. Sulla sostanza connettiva del cervello. Rendiconti dell' istitut Lombardo di scienze e lettere Fascicolo d'Aprile 1870.

1870. Gegenbaur. Grundzüge der vergleichenden Anatomie. Leipzig 1870.
Ph. Owsjannikow. Ueber das Nervensystem der Seesterne. Mélange biologiques T. VII 1871 p. 491—503. März 1870. (Bull. de l'acad. imp. de sciences de St. Pétersbourg. T. XV p. 310—318.)
— Histologische Studien über das Nervensystem der Mollusken. Vorläufige Mittheilung. Mélanges biologiques. T. VII. p. 689—685. Dec. 1870. (Bull. d. l'acad. d. sc. de St. Petersbourg T. XV p. 523—527.)
Eduard Brandt. Ueber das Nervenstystem der Lepas anatifera (anatomisch-histologische Untersuchung). Mélanges biologiques T. VII p. 504—515 jun. 1870. (Bull. d. l'acad. d. sc. St. Petersbourg T. XV p. 332—340.)

1871. Virchow. Die Cellularpathologie. 4 Aufl. Berlin 1871.
Jobert. Contribution à l'étude du système nerveux sensitif. Journal de l'anatomie et la physiol. par Robin. 1871.
Leydig. Ueber das Gehörorgan der Gastropoden. Arch. f. mikr. Anat. Bd. VII p. 202—119.
Max Schultze. Allgemeines über die Structurelemente des Nervensystems. Stricker Handb. d. Lehre v. d. Geweben. Leipzig. Bd. I p. 108—136.
C. Golgi. Contribuzione alla fina anatomia degli organi centrali del sistema nervoso. Rivista clinica di Bologna 1871—1872.

1872. August Solbrig. Ueber die feinere Structur der Nervenelemente bei den Gasteropoden. Gekr. Preisschr. Leipzig.
G. Huguenin. Neurologische Untersuchungen. 1. Ueber das Auge von Helix Pomatia. Zeits. f. wiss. Zool. Bd. XXII p. 126—136.
Nicolaus Kleinenberg. Hydra. Eine anatomisch-entwickelungsgeschichtliche Untersuchung. Leipzig.
Rindfleisch. Zur Kenntniss der Nervenendigungen in der Hirnrinde. Arch. f. mikr. Anat. Bd. VIII. p. 453—454.
Ranvier. Recherches sur l'histologie et la physiologie der nerfs. Arch. de phys. norm. et pathol. Paris 1872.
J. Gerlach. Ueber die Structure der grauen Substanz des menschlichen Grosshirns. Med. Centralblatt 1875 p. 273.
— Von dem Rückenmark. Stricker's Handbuch der Lehre v. d. Geweben p. 665—693.
Theodor Meynert. Vom Gehirne der Säugethiere. Stricker's Handb. der Lehre v. d. Gewebe p. 694—808.
Sigmund Mayer. Das sympathische Nervensystem. Stricker's Handbuch der Lehre v. d. Geweben p. 808—821.
Lacaze-Duthiers. Otocystes ou Capsules auditives des Mollusques (Gasteropodes). Arch. zool. exp. et gen. T. I p. 97—168.
— Du système nerveux des Mollusques gastéropodes pulmonés aquatiques. Arch. zool. exp. et gen. T. I p. 437—500.

1873. Edouard Claparède. Recherches sur la structure des Annelides sedentaire. Genève, Bale, Lyon. Nerv.syst. p. 112—131.

1873. P. Langerhaus. Untersuchungen über Petromyzon Planeri. Bericht über die Verhandl. der naturf. Gesellsch. zu Freiburg. Vol. VI. 1873.

C. Golgi. Sulla struttura della sostanza grigia del cervello. Communicazione preventiva. Gaz. med. Ital. Lomb. Ser. 6. T. VI.

Stieda. Studien über den Amphioxus lanceolatus. Mém. de l'Acad. des science de St. Petersb. Ser. 7, T. XIX No. 7. (Abstract in Jahresb. Anat. u. Phys. Bd. II 1875 p. 159.)

Sigm. Mayer. Zur Lehre von der Structur der Spinalganglien und der peripheren Nerven. Vorläufige Mitth. Wiener Akad. Anzeiger, 1873, p. 54.

Franz Boll. Die Histologie und Histiogenese der nervösen Centralorgane. Arch. f. Psychiatrie und Nervenkrankheiten. Bd. IV. Berlin.

1874. Fleischl. Ueber die Beschaffenheit des Axencylinders. Beitr. z. Anat. u. Phys. Festgabe an C. Ludwig 1874.

Auguste Forel. Les fourmis de la Suisse. Bale & Génève. Lyon.

Ludwig Stieda. Ueber den Bau der Cephalopoden. 1 Abth. Das centrale Nervensystem des Tintenfisches (Sepia officinalis). Zeits. f. wiss. Zool. Bd. XXIV p. 84—122.

H. D. Schmidt. Synopsis of the principal facts elicited from a series of microscopical researches upon the nervous tissues. Monthly microsc. journ. T. XII.

O. Bütschli. Beiträge zur Kenntniss des Nevensystems der Nematoden. Arch. f. mikr. Anat. Bd. X.

1875. Ernst Hermann. Das Central-Nervensystem von Hirudo medicinalis. Gekr. Preisschr. München.

Rabl-Rückhardt. Studien über Insectengehirne. Arch. f. Anat. u. Phys. 1875 p. 480—499.

Holl. Ueber den Bau der Spinalganglien. Sitzber. kais. Akad. Wiss. Wien. Bd. LXXII. p. 6.

L. Ranvier. Des tubes nerveux en T et de leurs relations avec les cellules ganglionnaires. Comptes rendus. Paris T. 81 p. 1274.

1876. F. M. Balfour. On the development of the spinal nerves in Elasmobranch. Fishes. Phil. trans. R. Soc. London. Vol. 166, p. 175—192.

Simroth. Ueber die Sinnesorgane unserer einheimischen Weichtiere. Zeits. f. wiss. Zool. Bd. XXVI p. 227—349.

C. Semper. Die Verwandtschaftsbeziehungen der gegliederten Thiere: III Strobilation und Segmentation. Arb. aus die zool.-zoot. Institut Würzburg. Bd. 3 p. 115.

Dionys Burger. Ueber das sogenannte Bauchgefaess der Lepidoptera, nebst einigen Betrachtungen über das sympatische Nervensystem dieser Insectenordnung. Niederländ. Arch. f. Zool. III, Heft. 2, p. 98—125.

P. Langerhaus. Zur Anatomie des Amphioxus lanceolatus. Arch. f. mikr. Anat. Bd. XII. Nerv.syst. p. 295—300.

Kuhnt. Die peripherische, markhaltige Nervenfaser. Arch. f. mikr. Anat. Bd. XIII p. 427.

1876. F. M. Balfour. On the spinal nerves of Amphioxus. Journ. of anat. and phys. Vol. X p. 689—692.
Axel Key und G. Retzius. Studien in der Anatomie des Nervensystems und Bindegewebes. Stockholm. 1875—76.
Thanhoffer. Zur Structur der Ganglienzellen der Invertebralknoten. Sitzb. d. k. ungarischen Akad. d. Wiss. Bd. VII.
Viault. Recherches histologiques sur la structure des centres nerveux des Plagiostomes. Arch. de zool. exp. et gen. Paris T. V.
Sigm. Mayer. Die periphere Nervenzelle und das sympatische Nervensystem. Arch. f. Psych. Bd. 6. p. 433.
Hensen. Beobachtungen über die Befruchtung und Entwichelung des Kaninchens und Meerschweinchens. Zeitschr. f. Anat. u. Entwickelungsgeschichte. B. I.
Schenk. Die Entwickelungsgeschichte der Ganglien und des Lobus electricus. Sitzber. d. kais. Akad. Wiss. Wien. Bd. LXXIV. p. 15.
v. Ihering. Die Gehörwerkzeuge der Mollusken in ihrer Bedeutung für das natürliche System derselben. Habilitationschr. Erlangen.
W. C. M'Intosch. On the central nervous system, the cephalic sacs, and other points in the anatomy of the Lineidæ. Journ. anat. & physiology. Vol. X p. 232—252.
— On the Arrangement and relations of the great nerve cords in the marine annelids. Proceed. of the Royal Soc. of Edinburgh. Session 1876—77.
Dietl. Die Organisation des Arthropodengehirns Zeitschr. f. wiss. Zool. Bd. XXVII p. 488—517.

1877. — Die Gewebselemente des Zentralnervensystems bei wirbellosen Thieren. Ber. d. naturw. med. Ver. Innsbruch 1877.
Franz Boll. Ueber Zersetzungsbilder des markhaltigen Nervenfasern. Arch. f. Anat. und Entwicklungsgesch. 1877.
H. v. Ihering. Vergleichende Anatomie und Physiologie des Nervensystems der Mollusken. Leipzig.
Thomas H. Huxley. A manual of the anatomy of inverbrated animals. London 1877.
Josef Victor Rohon. Das Centralorgan des Nervensystems der Selachier. Denkschr. d. math.-nat. Classe d. kais. Akad. d. Wiss. Wien. Bd. XXXVIII.
Sigm. Freud. Ueber den Ursprung der hinteren Nervenwurzeln im Rückenmark von Ammocoetes. Sitzb. d. kais. Akad. Wiss. Wien. Bd. LXXV.
Marshall. On the early stages of development of the nerves in birds. Jorn. of anat. and phys. Vol. XI, p. 491—515.
F. M. Balfour. The development of Elasmobranch fishes. Journ. of anat. and phys. Vol. XI. Part III (1877), p. 438—439.
Denissenko. Zur Frage über den Bau der Kleinhirnrinde bei verschiedenen Classen von Wirbelthieren. Arch. f. mikr. Anat. Bd. XIV.

1878. Hans Schultze. Axencylinder und Ganglienzelle. Mikroskopische Studien über die Structur der Nervenfaser und Nervenzelle bei Wirbelthieren. Archiv für Anatomi und Entwickelungsgesch. Leipzig 1878 p. 259—287.

E. Berger. Untersuchungen über den Bau des Gehirns und der Retina der Arthropoden. Arbeiten aus dem Zool. Inst. der Universität. Wien. T. I Heft. 2 1878 p. 1—48.

— Nachtrag zu den Untersuch etc. ibid. Heft. 3 p. 1—5.

Floegel. Ueber den einheitlichen Bau des Gehirns in den verschiedenen Insektenordnungen. Zeitschr. f. wiss. Zool. Bd. 30, Suppl. p. 556—592. 1878.

Émile Yung. Recherches sur la Structure intime et les fonctions du système nerveux central chez les crustacés Décapodes. Arch. de zool. expér. et gen. T. 7, p. 401—534.

— Compt. rendus. Paris. T. 88 (Febr. 1879), p. 240—242.

Fr. Spangenberg. Bemerkungen zur Anatomi der Limnadia Hermanni, Brongn. Festschrift zur Feier des Fünfzigjähr. Doctorjub. am 22 April 1878 Herrn Prof. von Silbold gew. Also in: Zeits. f. wiss. Zool. Bd. 30. Suppl. Nerv. syst. p. 480—491.

G. Bellonci. Morfologio del sistema centrale nervoso della Squilla Mantis. Annali de museo civico di storia naturale di Genova. Vol. XII p. 518—545.

Dietl. Untersuchungen über die Organisation des Gehirns wirbelloser Thiere. Sitzber. der kais. Akad. Wiss. Wien. Bd. LXXVII. 1 Abth. p. 481—532, 584—603.

Cadiat. Note sur la structure des nerfs chez les Invertebrés. Comptes rendus. Paris. T. 86. (3 Juin), p. 1420—1422.

Sigm. Freud. Ueber Spinalganglien und Rückenmark des Petromyzon. Sitzungsber. d. kais. Akad. Wiss. Wien. LXXVIII. 3 Abth. Juli Heft. 1878 p. 81—167.

Fritsch. Untersuchungen über den feineren Bau des Fischgehirns. Berlin 1878.

A. Sanders. Contributions to the anatomy of the central nervous system in vertebrate animals. Phil. transact. Royal Society London (1878). Vol. 169 P. II. London 1879.

Ranvier. Leçons sur l'histologie du systéme nerveux. Paris.

— Traité technique d'histologie. Paris.

O. und R. Hertwig. Das Nervensystem und die Sinnesorgane der Medusen. Leipzig.

A. Zincone. Sulla prominenze de midolo spinale delle Trigle. Napoli.

G. Bellonci. Richerche intorno all' intima tessitura del cervello dei Teleostei. Memorie della R. Accad. dei Lincei. Anno CCLXXVI (1878—1879). Roma.

1878. Rich. Krieger. Ueber das centrale Nervensystem des Flusskrebses. Zool. Anzeiger I Jahr. No. 15 p. 340—342.
— Ueber das Centralnervensystem des Flusskrebses. Dissert. Leipzig 1879.

1879. Hans Schultze. Die fibrilläre Structur der Nervenelemente bei Wirbellosen. Arch. f. mikr. Anat. Bd. XVI p. 57—111.

O. und R. Hertwig. Die Actinien anatomisch und histologisch mit besonderer Berücksichtigung des Nervenmuskelsystems untersucht. Jenaische Zeits. f. Naturwiss. Bd. 13 p. 457—640 (1879); Bd. 14 p. 39—89 (1880).

Josef Victor Rohon. Untersuchungen über den Bau eines Microcephalen-Hirnes. Arb. aus den zool. Instit. der Univ. Wien und d. zool. Stat. in Triest. Tom II 1 Heft. p. 1—58 1879.

C. Claus. Der Organismus der Phronimiden. Arbeiten aus der zool. Inst d. Universität Wien u. zool. Stat. Triest. T. II Heft. 1. Nerv. syst. p. 43—75.

Stricker und Unger. Untersuchungen über den feineren Bau der Grosshirnrinde. Sitzber. der kais. Akad. Wiss. Wien. 1879.

Arndt. Etwas über die Axencylinder der Nervenfasern. Virchow's Archiv Bd. LXXVIII.

E. T. Newton. On the brain of the cockroach, Blatta orientalis. Quart. journ. micr. science London. Vol. 19 p. 340—356.

Arnold. Lang. Untersuchungen zur vergleichenden Anatomie und Histologie des Nervensystems der Plathelminthen. I. Das Nervensystem der marinen Dendrocoelen. Mitth. zool. Stat. Neapel. Bd. 1. p. 460—488.

J. v. Kennel. Die in Deutschland gefundenen Landplanarien, Rhynchodemus terrestris, O. F. Müller, und Geodesmus bilineatus, Mecznikoff. Arb. zool. zoot. Institut in Würzburg. Bd. V. 1882 p. 149—158.

E. O. Taschenberg. Beiträge zur Kenntniss ectoparositischer mariner Trematoden. Halle 1879 (Besonders abgedruckt aus der Abh. der Naturf. Gesellsch. in Halle. Bd. XIV, 3) Nerv. syst. p. 16—20.

Ed. Brandt. Vergleichend-anatomische Skizze der Nervensystems der Insekten. Horae societatis entomologicae Rossicae T. XV.

Anton Schneider. Beiträge zur Vergleichenden Anatomie und Entwickelungsgeschichte der Wirbelthiere. Amphioxus lanceolatus. Nervensystem p. 12—17. Berlin.

G. Bellonci. Richerche comparative sui centri nervosi dei Vertebrati. Memorie della r. accad. d. Lincei. anno CCLXXVII (1879, 1880). Roma.

R. Wiedersheim. Ueber das Gehirn und die spinalartigen Hirnnerven von Ammocoetes. Zool. Anzeiger. II Jahrg. p. 589—592.

1880. R. Wiedersheim. Das Gehirn von Ammocoetes und Petromyzon Planeri mit besonderer Berücksichtigung der spinalartigen Hirnnerven. Jenaische Zeits. f. Naturwiss. Bd. 14, p. 1—24.

1880. Oscar Hertwig. Die Chaetognathen. Eine Monographie. Jen. Zeitschr. f. Nat. Bd. 14. Nerv. syst. p. 223—236.

A. A. W. Hubrecht. Zur Anatomie und Physiologie des Nervensystemes der Nemertinen. Verh. van de Koninkl. Akad. van Wetensch. Amsterdam Vol. XX.

Karl Richard Krieger. Ueber das Centralnervensystem des Flusskrebses. Zeitschr. f. wiss. Zool. Bd. 33 p. 527—594. 1880.

H. Michels. Beschreibung des Nervensystems von Oryctes nasicornis im Larven-, Puppen- und Käferzustande. Zeitsch. f. wiss. Zool. Bd. 34 p. 641—702. Sept. 1880.

G. Retzius. Untersuchungen über die Nervenzellen der cerebrospinalen Ganglien und der übrigen peripherischen Kopfganglien, mit besonderer Rücksicht auf die Zellenausläufer. Arch. f. Anat. u. Entwickelungsgeschichte 1880 p. 369.

Stiénon. Recherches sur la structure des ganglions spinaux chez les Vertebrés suppérieurs. Annales de l'université libre de Bruxelles 1880.

G. Bellonci. Ricerche comparative sulla struttura dei centri nervosi dei vertebrati. Memorie della. r. accad. dei Lincei. Roma. Vol. V. Seduta 1 Febbraio.

— Sistema nervoso e organi dei sensi dello Sphæroma serratum. R. accademia dei Lincei, serie 3. vol. X 1880—1881. (Systéme nerveux et organes des sens du sphæroma serratum. Arch. ital. de biologie. T. I, p. 176—192. 1882.)

— Sui lobi olfativi del Nephrops norvegicus. Mem. dell' accademia delle scienze di Bologna. 1880.

1881. — Contribuzione alla istologia del cervelletto. Accademia dei Lincei. Gennaio 1881 p. 45—49.

— Ueber den Ursprung des Nervus opticus und den feineren Bau des Tectum opticum der Knochenfische. Zeits. f. wiss. Zool. Bd. 35, p. 23—29.

Arnold Lang. Untersuchung etc. II. Ueber das Nervensystem der Trematoden. Mitth zool. Stat. Neapel. Bd. 2, p. 28—52. 1881, and

— III. Das Nervensystem der Cestoden im Algemeinen und dasjenige der Tetrarynchen im Besondern. ibid. p. 372—400. 1881.

— Untersuchungen zur etc. IV. Das Nervensystem der Tricladen. V. Vergleichende Anatomie des Nervensystems der Plathelminthen. Mitth. zool. Stat. Neapel. Bd. 3, p. 53—96. Decemb. 1881.

N. Kleinenberg. Sull' origine del sistema nervoso centrale degli anellidi. Reale accademia dei Lincei anno CCLXXVIII (1880—81). Roma.

— De l'origine du systeme nerveux central des Annelides. Arch. ital. de biologie T. I p. 62—77. 1882.

Charles Julin. Recherches sur l'organisation des Ascidies simples. Arch. de biologie. T. II p. 59—126.

1881. A. Gerstaecker. Gliederfüssler: Arthropoda. Bronn's Klassen und Ordnungen des Thier-Reiches. Bd. V, II Abth. Isopoda. Nervensystem p. 44—55. Amphipoda. Nervensystem p. 326—341.

Ray Lankester. Observations and reflections on the appendages and on the Nervous system of Apus cancriformis. Quart. journ. micr. sci. Vol. XXI. Nerv. syst. p. 367—376.

Jos. Th. Cattie. Beiträge zur Kenntnis der Chorda supra-spinalis der Lepidoptera und des centralen, peripherischen und sympathischen Nervensystems der Raupen. Zeits. f. wiss. Zool. Bd. 35 p. 304—320. 1881.

J. W. Spengel. Geruchsorgan und Nervensystem der Mollusken. Zeitschr. f. wiss. Zool. Bd. 35 p. 333—383.

— Oligognatus Bonelliae, eine schmarotzende Eunicee. Mittheil. aus d. zool. Stat. Neapel. Bd. 3 Nerv.syst. p. 27—47.

P. Mayser. Vergleichend-anatomische Studien über das Gehirn der Knochenfische mit besonderer Berücksichtigung der Cyprinoiden. Zeits. f. wiss. Zool. Bd. XXXVI (1882) p. 259—364.

W. Vignal. Note sur l'anatomie des centres nerveux du Mola. Arch. de zool. exp. et. gen. T. IX p. 369—386.

G. Schwalbe. Handbuch der Neurologie. Erlangen 1881.

Camillo Golgi. Studi istologici sul midollo spinale. Archivio italiano per le malattie nervose. 1881 p. 155—165.

— Sulla origine centrale dei nervi. Giornale internazionale 1881, p. 15.

1882. Camillo Golgi. Origine du tractus olfactorius et structure des lobes olfactifs, de l'homme et d'autres mammifères. Arch. ital. de biologie T. I p. 454—462.

— Considérations anatomiques sur la doctrine des localisations cérébrales. Arch. ital. de biologie. T. II p. 237—253; p. 255—268. Abrstract from: Gazzetta degli ospitali, anno III, N. 61, 62, 63, 64 67, 69, 70, 71, 72.

G. B. Laura. Sulla struttura del midollo spinale. Atti della r. accad. di medicina di Torino. 1882 p. 1—89.

— Sur la structure de la moelle épinière. Arch. ital. de biologie. T. I p. 147—175. 1882.

Sigm. Freud. Ueber den Bau der Nervenfasern und Nervenzellen beim Flusskrebs (15 Dec. 1881). Sitzber der kais. Akad. Wiss. Wien. Bd. LXXXV. 3 Abth. p. 9—46.

Eduard Meyer. Zur Anatomie und Histologie von Polyophthalmus pictus, Clap. Arch. f. mikr. Anat. Bd. XXI. Nerv. syst. p. 782—790.

Josef Victor Rohon. Ueber den Ursprung des Nervus acusticus bei Petromyzonten. Sitzber. der kais. Akad. Wiss. Wien. Bd. LXXXV. 1 Abth. Jahrg. 1882, p. 245—267.

1882. M. Ussow. De la structure des lobes accessoires de la moelle épinière de quelques poissons osseux. Arch. d. biologie. T. III p. 605—658.

Berhard Rawitz. Ueber den Bau der Spinalganglien. Arch. f. mikr. Anat. Bd. XXI p. 244—290.

W. Flemming. Vom Bau der Spinalganglienzellen. — In: Beiträge zur Anatomie und Embryologie als Festgabe für J. Henle. Bonn.

L. Ranvier. Sur les ganglions cérébro-spinaux. Comptes rendus. Paris 1882 p. 1167.

1883. W. Vignal. Recherches histologiques sur les centres nerveux de quelques Invertébrés. Arch. zool. exq. et gen. Ser. 2 T. I. p. 267—412.

Ludvig Böhmig. Beiträge sur Kenntniss des Centralnervensystems einiger pulmonaten Gasteropoden. Inaug. Disert. Leipzig.

Köstler. Eingeweidenerveasystem von Periplaneta orientalis. Zeitschr. f. wiss. Zool. Bd. 39 p. 572—595.

Rabl-Rückhard. Grosshirn der Knochenfische und seine Anhangs-gebilde. Arch. f. Anat. u. Entwick. 1883.

Romeo Fusari. Sull' origine delle fibre nervoso nello strato moleculare delle circonvoluzioni cerebellari dell' uomo. Atti della r. accademia delle scienze di Torino. Vol. XIX.

Camillo Golgi. Recherches sur l'histologie des centres nerveux. Arch. ital. de biologie. T. III p. 285—317 (1883); T. IV p. 92—123 (1883); T. VII p. 15—47 (1886).

Ph. Owsjannikow. Ueber das sympathische Nervensystem des Flussneunauges, nebst einigen histologischen Notizen über andere Gewebe desselben Thieres. Mélanges biol. des Bullet. ac. imp. des sci. St. Petersbourg T. XI.

Leydig. Untersuchungen zur Anatomie und Histologie der Thiere. Bonn 1883.

Friedrich Ahlborn. Untersuchungen über das Gehirn der Petromyzonten. Zeits. f. wiss. Zool. Bd. 39 p. 491.

— (Ueber den Ursprung und Austritt der Hirnnerven von Petromyzon. ibid Bd. 40 p. 286—308. 1884).

Émile Baudelot. Recherches sur le système nerveux des Poisons. Paris.

W. Ainslie Hollis. Researches into the histology of the central grey substance of the spinal cord and medulla oblongata. Journ. of Anat. and phys. Vol. XVII, p. 517—522; XVIII, p. 62—65; p. 203—207, p. 411—415.

G. Bellonci. Sur la structure et les rapports des lobes olfactifs dans les Arthropodes supérieurs et les Vertébrés. Arch. ital. de biologie. T. III, p 191—196. 1883.

— Les lobes optiques des oiseaux. Note préliminaires. Arch. ital. de biologie. T. IV, p. 21—26.

1884. G. Bellonci. La termination centrale du nerf optique chez les Mammifères. Arch. ital. de biologie. T. VI p. 405—412. — Also in: Mem. d. r. accad. d. sci. Bologna. Ser. 4; T. VI, p. 199—204. 1885.

Wladimir Schimkewitsch. Étude sur l'anatomie de l'épeire. Ann. d. sci. nat. Ser. 6. T. XVII. Nerv.syst. p. 15—31. (April 1883.)

H. Viallanes. Études histologiques et organologiques sur les centres nerveux et les organes des sens des animaux articulés. Premier mémoire. Le ganglion optiques de la Langouste (Palinurus vulgaris). Ann. d. sci. nat. Ser. 6. T. XVII.

— Études histologiques etc. Deuxième mémoire. Le ganglion optique de la Libellule (Æschna maculatissima) ibid. T. XVIII.

Ed. van Beneden & Ch. Julin. Le système nerveux central des Ascidies adultes et ses rapports avec celui des larves urodèles. Arch. de biologie. T. V.

Franz Vejdovský. System und Morphologie der Oligochaeten. Prag. Nervensystem p. 79—96.

Arnold Lang. Die Polycloden des Golfes von Neapel etc. Fauna und Flora d. Golfes v. Neapel. (Zool. Stat. Neapel.) XI Monogr. Nerv. syst. p. 164—190. Histologie. p. 182.

A. S. Packard. On the structure of the brain of the sessileeyed Crustacea. National academy of sciences vol. III. Washington Apr. 1884.

Julien Fraipont. Recherches sur le systeme nerveux central et périphérique des Archiannelides etc. Archives de biol. T. V. p. 243—304.

Remy Saint-Loup. Recherches sur l'organisation des Hirudinées. Ann. d. sci. nat. Ser. 6. T. XVIII. Nerv.syst. p. 47—63.

Alexander Foettinger. Recherches sur l'organisation de Histriobdella homari, P. J. van Beneden, rapportée aux Archiannelides. Arch. de biologie. T. V. Nerv.syst. p. 445—453.

Livio Vincenzi. Note histologique sur l'origine réelle de quelques nerfs cérébraux. Arch. ital. de biologie. T. V. p. 109—130.

Sigm. Freud. Eine neue Methode zum Studium des Faserverlaufs im Centralnervensystem. Arch. f. Anat. u. Entwicklungsgesch. 1884. p. 453—460.

W. S. Colman. Notes on the minute structure of the spinal cord of a human foetus. Journ. of anat. & phys. Vol. XVIII, p. 436—441.

A. D. Ónodi. Ueber die Entwickelung der Spinalganglien und der Nervenwurzeln. Internat. Monatschr. f. Anat. u. Hist. Bd. I, Heft. 3 & 4 (1884).

1885. G. Pruvot. Recherches anatomiques et morphologiques sur les système nerveux des Annélides Polychètes. Arch. zool. exp. et gen. Paris. Ser. 2. T. III p. 211—336.

E. Rohde. Beiträge zur Kennt uns der Anatomie der Nematoden. Zoologische Beiträge von A. Schreider. Breslau. Bd. I. Nerv.syst. p. 12—18.

1885. Franz Leydig. Zelle und Gewebe. Bonn. p. 5—7 und 164—209.
J. Poirier. Contribution à l'histoire des Trématodes. Arch. zool. expér et gén. Paris. Ser. 2. T. III. Nerv.syst. p. 598—613.
Hans Gierke. Die Stützsubstanz des Centralnervensystems. Arch. f. mikr. Anat. Bd. XXV, p. 441—554 (1885); Bd. XXVI, p. 129—228 (1886).
A. D. Onodi. Ueber die Gangliengruppen der hinteren und vorderen Nervenwurzeln. Centralblatt f. die medicinischen Wissenschaften. 1885, Nr. 16 und 17.
Livio Vincenzi. Sulla morfologia cellulare del midollo allungato e istmo dell'encefalo. Memorie della r. accad. d. scienze d. Torino. Ser. 2 Tom. XXXVII.
— Sull' origine reale del nervo ipoglosso. Atti della r. Accad. delle scienze de Torino. Vol. XX, p. 798—806.
— Sull' origine reale del nervo pneumogastrico. Comunic. preventiva. Gazz. delle cliniche. Vol. XXI, p. 209—211.
A. A. Torre. Sulla cariocinesi nel tessuto nervoso. Gazz. delle cliniche. Vol. XXI, p. 282.
Camillo Golgi. Sulla fina anatomia degli organi centrali del sistema nervoso. Milano.
Béla Haller. Untersuchungen über marine Rhipidoglossen. II Textur des Centralnervensystemes und seiner Hüllen. Morph. Jahresbuch. Bd. XI, p. 321—436. 1885—1886.
Fridtjof Nansen. Bidrag til Myzostomernes Anatomi og Histologi. Bergens Museum. Bergen 1885. Nerv-syst. p. 10—40, 71—75.

1886. Franz von Wagner. Das Nervensystem von Myzostoma (F. S. Leuckart). Graz. (52 p.)
A. D. Ónodi. Ueber die Entwickelung des sympatischen Nervensystems. Arch. f. mikr. Anat. Bd. XXVI, p. 61—80, 553—590.
Fridtjof Nansen. Foreløbig Meddelelse om Undersøgelser over Centralnervesystemets histologiske Bygning hos Ascidierne samt hos Myxine glutinosa. Bergens Museum's Aarsberetning for 1885.
— Preliminary Communication on some investigations upon the histological structure of the central nervous system in the Ascidia and in Myxine glutinosa. Ann. mag. nat. hist. London. Ser. 5. Vol. XVIII p. 209—226.
J. Niemic. Untersuchungen über das Nervensystem der Cestoden. Arb. a. d. zool. Inst. Wien u. d. zool. Stat. Triest. Tom. VII. Heft. 1 p. 1—60.
Gustav Fritsch. Ueber einige bemerkenswerte Elemente des Centralnervensystems von Lophius piscatorius, L. Arch. f. mikr. Anat. Bd. XXVII p. 13—31.
F. H. Herrick. Notes on the embryology of Alpheus and other Crustacea and on the development of the compound eye. John Hopkins university circulars. Vol. VI, No. 54, p. 42—44. Baltimore.

1887. W. B. Ransom and D'Arcy W. Thompson. On the spinal and visceral nerves of Cyclostomata. Zool. Anzeiger. IX Jahrg., p. 421—426. 1886.

Nicolaus Kleinenberg. Die Entstehung des Annelids aus der Larve von Lopadorhynchus. Zeits. f. wiss. Zool. Bd. XLIV. Neuromuskesystem p. 58—151.

Emil Rohde. Histologische Untersuchungen über das Nervensystem der Chaetopoden. Sitzber. d. k. pr. Akadem. der Wiss. Berlin. XXXIX p. 781—786. Juli 1886. Translated in: Ann. mag. nat. hist. Ser. 5. Vol. XVIII p. 311—316. 1886.

Michael v. Lenhossék. Untersuchungen über die Spinalganglien des Fsosches. Arch. f. mikr. Anat. Bd. XXVI, p. 370—453.

G. Fritsch. Uebersicht der Ergebnisse einer anatomischen Untersuchung über den Zitterwels (Malopterus electricus). Sitzungsber, der k. preus. Akad. der Wissensch. Berlin. XLIX. L. 2 Decemb. 1886, p. 1137—1140.

F. Leydig. Die riesigen Nervenröhren im Bauchmark der Ringelwürmer. Zool. Anzeiger No. 234. 1886.

Romeo Fusari. Richerche intorno alla fina anatomia dell' encefalo dei Teleostei. Nota preventiva. Bollettino scientifico N. 2, Giugno 1886. Pavia.

1887. Béla Haller. Ueber die sogenannte Leydig'sche Punktsubstanz im Centralnervensystem. Morph. Jahresbuch. Bd. XII. 1887. p. 325—332.

During the printing of the present memoir the following papers of interest to our subject have appeared:

1887. Bernhard Rawitz. Das centrale Nervensystem der Acephalen. Jenaische Zeitschr. f. Naturwissenschaft. Bd. 20, p. 386—461.

Julius Waldschmidt. Zur Anatomie des Nervensystems der Gymnophionen. Jenaische Zeitschr. f. Naturwiss. Bd. 20, p. 461—476. 1887.

L. Edinger. Vergleichend-entwicklungsgeschichtliche Studien im Bereich der Gehirn-Anatomie. Anatom. Anzeiger. Bd. II. No. 6, p. 145—143. (März 1887.)

H. Viallanes. Études hist. etc. Quatrième mémoire. Le cerveau de la Guêpe (Vespa crabro et V. vulgaris). Ann. sci. nat. Ser. 7. T. 2.

Alfred Sanders. Contributions to the anatomy of the central nervous system in vertebrate animals (Plagiostomata). Phil. trans. Royal. Soc. London. Vol. 177. Part II, p. 733—766. (Read. Jan. 1886.)

A. A. W. Hubrecht. Report on the Nemertea collected by H. M. S. Challenger during the years 1873—76. Zool. Chall. Exp. Part. LIV. 1886 vol. XIX. Nerv.syst. p. 73—90.

W. Bechterew. Le cerveau de l'homme dans ses rapports et connexions intimes. Arch. slaves de biologie. T. III, Fasc. 3, p. 293—321 (Mai); T. IV, Fasc. 1, p. 1—30 (Juillet).

1887. Willy Kükenthal. Ueber das Nervensystem der Opheliaceen. Jenaische Zeits. f. Naturwiss. Bd. XX, p. 511—580.

Michael v. Lenhossék. Beobachtungen am Gehirn des Menschen. Anat. Anzeiger. Bd. II. No. 14, p. 450—461. Juni.

A. Koelliker. Die Untersuchungen von Golgi über den feineren Bau des zentralen Nervensystems. Anat. Anzeiger. Bd. II. No. 15, p. 480—483 (Juli).

Charles Julin. Le système nerveux grand sympathique de l'Ammocoetes (Petromyzon Planeri). Communication préliminaire. Anat. Anzeiger II. Jahrg. Nr. 7, p. 192—201. 1887.

Romeo Fusari. Untersuchungen über die feinere Anatomie des Gehirnes der Teleostier. Intern. Monatschr. f. Anat. u. Phys. Bd. IV, p. 275—300.[1]

---

[1] Last winter I wrote a paper: Über das Nervensystem der Myzostomen, which was going to be published in Jenaische Zeitschr. f. Naturwiss. but which is not yet appeared as far as I know. As that paper was written before the investigations, above described, had been finished, there are some remarks on the nervous system of various animals which do not quite agree with the results we have here obtained.

Explanation of the plates.

Plate I.

The illustrations are drawn under the *camera lucida*, from the microscope directly upon the stone. The preparations (exept fig. 5 & 6) were' fixed i chromo-aceto-osmic acid and stained with hæmatoxylin.

Fig. 1. *Homarus vulgaris*. (Magnified 60 diameters; Zeiss AA, 2.) Transverse sections of an oesophageal commissure. *a* External sheath. *b* Layer of connective tissue inside the ext. sheath. *c* Inner layer or sheath of connective-tissue closely applied to the contents of nerve-tubes (compare fig. 5, *a*). t, t', t_1 Large nerve-tubes with more or less concentrated axes. t'' Large nerve-tubes with no axis. *nt* Central bundle of largish nerve-tubes. *s nt* Masses of small nerve-tubes, more peripherically situated.

» 2. *Homarus vulgaris*. (Magnified 950 diameters; Zeiss. Hom. im. $1/_{18}$, 1.) A part of fig. 1, representing some smallish nerve-tubes of various sizes, more highly magnified. The transsected primitive tubes, of which the contents of the nerve-tubes are composed, are distinctly visible as round meshes. t, t Largish nerve-tubes in which a concentration towards an axis is visible. t', t' Small nerve-tubes; in one of them a slight concentration towards an axis is visible. t'' Small nerve-tube in which no concentration towards an axis is visible. *a* Vacuoles. *b* Neuroglia-substance. *c* The sheath of a large nerve-tube which is illustrated in fig. 1, t_1. *d* Vacuoles in some nerve-tubes, probably artificially produced. *k* Neuroglia-nuclei.

» 3. *Homarus vulgaris*. (Magnified 950 diameters; Zeiss. Hom. im. $1/_{18}$, 1.) Transverse section of a large nerve-tube with axis (it is the same tube which is illustrated in fig. 1, t'). *a* Neuroglia-sheath. *b* Sheath of another large nerve-tube.

Fig. 4. *Homarus vulgaris* (Magnified 950 diameters; Zeiss. Hom. im. $1/_{18}$, 1). Longitudinal section of the same nerve-tube as is illustrated in fig. 3 (of the same piece of an oesophageal commissure was first taken transverse sections, and subsequently longitudinal sections. *a, a'* Deeply staining axis. *b* External layers of primitive tubes which are somewhat obliquely transsected. *c & d* Neroglia-sheath of the nerve-tube, which on one side (*c*) is considerably thicker than on the other. *k* Neuroglia-nucleus situated inside the sheath.

» 5. *Nephrops norvegicus*. (Magnified 950 diameters; Zeiss. Hom. im. $1/_{18}$, 1.) Transverse section of nerve-tubes, from an oesophageal commissure. *a* Inner layer of the connective-tissue surrounding the bundle of nerve-tubes of the commissure (compare fig. 1, *c*). *b* Outer layer of the same connective-tissue (compare fig. 1, *b*). *t, t* Transsected nerve-tubes; in the right one is a very slight concentration towards a deeply staining axis visible. *t', t'', t'''* Nerve-tubes with vacuoles, which possibly, to some extent, are transsected tubes. t_1 Nerve-tubes containing solely large meshes, which possibly are transsected tubes. *e* Vacuoles. *k, k_1, k', k'', k'''* Neurogla-nuclei and nuclei of the connective-tissue. *k'* Neuroglia-nucleus situated inside the sheath of a nerve-tube. The preparation was fixed in *Lang's fluid* (corrossive sublimat 12 %) and stained with *hæmotoxylin*.

» 6. *Nephrops norvegicus*. (Magnified 1700 diameters; Zeiss. Hom. im. $1/_{18}$, 3.) Part of the contents of a nerve-tube highly magnified to show the walls of the primitive tubes. *a* Thickenings in these walls along the concreting edges of the tubes. From the same preparations as fig. 5.

» 7. *Homarus vulgaris*. (Magnified 190 diameters; Zeiss CC, 2). Transverse section of a small *peripheral nerve* containing large and small nerve-tubes. *t* Large nerve-tubes; their sheaths are very thick and consist of several layers. *n t* Small nerve-tubes. *s* Septa, dividing the small nerve-tubes into bundles. *k* Neuroglia- and connective-tissue-nuclei.

» 8. *Homarus vulgaris*. (Magnified 1150 diameters; Zeiss. Hom. im. $1/_{18}$, 2.) Transsected nerve-tubes from a peripheral nerve. *nt* Small nerve-tubes containing some few primitive tubes only. *p, p* Thickenings along the concreting edges of the sheaths of the small nerve-tubes. *t* Large nerve-tubes outside the scheath of which a neuroglia-nucleus is situated.

Fig. 9. *Nereis virens.* (Magnified 950 diameters; Zeiss $^1/_{18}$, 1.) Longitudinal section of a large nerve-tube (*b*) and some small nerve-tubes (*nt*) from the ventral nerve-cord. *a* Sheath of the large nerve-tube (*b*). The spongioplasmic walls of the primitive tubes of the nerve-tubes are distinctly seen longitudinal transsected.

» 10. *Nereis virens.* (Magnified 950 diameters; Zeiss $^1/_{18}$, 1.) Transverse sections of some large nerve-tubes running centrally along the ventral nerve-cord (compare fig. 14). *a* Central fibrous septum dividing the nerve-cord longitudinally into two lateral cords (cf. fig. 14). *b* & *b'* Transverse bundles of neuroglia-fibres; subdividing into branches (*c, c*). *t, t* Large central nerve-tubes. t_1, t_2, t_3 Somewhat smaller nerve-tubes, more ventrally situated. *nt* Small nerve-tubes. *nt'* Small nerve-tubes situated between the large central nerve-tubes (*t*) and the central septum (and the transverse bundles of fibres [*b*]). *k* Neuroglia-nucleus situated in the ventral, reticular part (B) of the central septum (*a*).

» 11. *Lumbricus agricola.* (Magnified 500 diameters; Seibert $^1/_8$, Zeiss 2.) Transverse section of the three colossal nerve-tubes (t_1, t, t) and a part of the dotted substance (*c*). *a* Connective-tissue outside the perineurium. *b* Perineurium (Vignal's gaine troisiéme), outside which the layer of muscles (*m, m*) is situated. *d* Neuroglia extending inside the perineurium, between the three colossal nerve-tubes. *s, s, s* Septa or very thick fibres occurring in this neuroglia and connected with the perineurium. *c* Dotted substance, exhibiting a reticulation produced by a transsection of tubes. *nt* Transsected nerve-tubes. *k* Neuroglia-nuclei.

Plate II.

Fig. 12. *Homarus vulgaris.* (Magnified 330 diameters, Zeiss. Apochr. 4.0, — 0.45, 1. Cam. luc.) Large nerve-tubes from an oesophageal commissure, seen in the live state in the commissure.

A & B Large nerve-tubes with striated axes (*c, d*). *a, a* Neuroglia layers between the nerve-tubes. *b* Small nerve-tube with a visible longitudinal striation. *e, e* Small nerve-tubes without visible striation.

Fig. 13. *Homarus vulgaris*. (Magnified 950 diameters; Zeiss. Hom. im. $^1/_{18}$, 1. Cam. luc.) Large nerve-tubes (*a, a*) from a peripheral, abdominal nerve, and seen in the nerve in the live state; no striation is visible. *b* Sheaths of the nerve-tubes. *c, c, c'* Apparent septa, optic illusions produced by a slight bending of the nerve-tubes and their sheaths. *k* Neuroglia-nucleus situated inside the sheath of the one nerve-tube.

» 14. *Nereis virens*. (Magnified 190 diameters; Zeiss. CC. 2. Cam. luc. direct upon the stone.) Transverse section of the ventral nerve-cord. *cu* External cuticulum. *ep* Thick ectodermal epithelium, outside the ventral nerve-cord. *k* Nuclei of the high cylindrical cells of this ectoderm. ep_1 The common ectoderm of the body. k_1 The nuclei of this ectoderm. *n, sh* »Connective« or neuroglia tissue surrounding and enveloping the ventral nerve-cord. *k' k''* Nuclei of this tissue. k_1 Similar nuclei, situated more ventrally, towards the ectodermal epithelium, from the nuclei of which they are not easily distinguished. *gc* & *gc'* Ganglion cells. *m, m* Membrane separating the ectoderm and the ventral nerve-cord from the inner layers of the body. (From a preparation fixed in Flemming's chromo-aceto-osmic acid and stained with Delafield's hæmatoxylin.)

» 15. *Homarus vulgaris*. (Magnified 620 diameters; Zeiss F, 1. Cam. luc.) Isolated, largish nerve-tubes from an oesophageal commissure, macerated in acetic acid (1), glycerin (1) and aqua dest. (5).

» 16. *Homarus vulgaris*. (Magnified 620 diameters; Zeiss F, 1. Cam. luc.) Isolated smallish nerve-tubes from an oesophageal commissure, macerated in the same way as in fig. 15.

» 17. *Patella vulgata*. (Magnified 620 diameters; Zeiss F, 1. Cam. luc. direct in the stone.) Nerve-tubes from a pedal nerve-cord and a peripheral nerve, macerated as above.

» 18. *Patella vulgata*. (Magnified 620 diameters; Zeiss F, 1. Cam. luc.) Nerve-tubes macerated as above, stained with picro-carmine.

Fig. 19. *Patella vulgata.* (Magnified 620 diameters; Zeiss F, 1. Cam. luc. direct upon the stone.) Transverse section of a nerve, from which two smaller nerves (a and b) laterally issue. v Vental side of the nerve. a & b Transversally transsected nerve-tubes of the small lateral nerves. a' & b' Longitudinally transsected nerve-tubes of the same nerves. c & d Large nerve-tubes, transversally transsected s, s Septa dividing the nerve into bundles of nerve-tubes. s', s' Septa separating the nerve-tubes of the two small lateral nerves from the rest of the large nerve. f Neuroglia-cell with nucleus situated inside the neurilem-sheath k Neuroglia-nuclei. (From a preparation treated with osmic acid (1 %) for 48 hours, stained with Delafield's hæmatoxylin and decoloured in water to wich some drops of acetic acid had been added; the section was taken directly with a razor without imbedding.)

» 19 A. A part of the section, illustration in fig. 19, highly magnified to show the structure of the walls between the nerve-tubes.

» 20. *Patella vulgata.* (Magnified 750 diameters; Zeiss F, 2. Cam. luc. direct upon the stone.) Longitudinal section of a peripheral nerve near its origin in the pedal nerve-cord. gc A ganglion cell sending a protoplasmic process (p) towards the neurilem-sheath (sh); the nervous process which runs longitudinally was not distinctly visible in this section. k, k', k'' Circular and oblong, deeply staining neuroglia-nuclei. n, nn, nn', nn'', nn''' Ovoid, less staining and, generally, larger neuroglia-nuclei. (The preparation was fixed in chromo-aceto-osmic acid and stained with Delafield's hæmatoxylin.)

» 21. *Phallusia obliqua.* (Magnified 350 diameters; Zeiss CC, 4. Cam. luc. direct upon the stone.) Transverse section of a peripheral nerve. sh External neurilem-sheath. (From a preparation fixed and hardened in Müller's fluid, stained with acid fuchsin dissolved in turpentine with a little abs. alcohol added.)

» 22. *Phallusia obliqua.* (Magnified 750 diameters; Zeiss F, 2. Cam. luc.; drawn directly upon the stone.) Longitudinal section of a part of a peripheral nerve. a & b Septa (longitudinally transsected) between bundles of nerve-tubes. n Neuroglia-nucleus situated in one septum (a). ts Sheaths of the nerve-tubes. The spongioplasmic walls of the primitive tubes are distinctly seen in the nerve-tubes. (From a preparation treated as above.)

Plate III.

The illustrations of this plate are drawn under the *camera lucida*, from the microscope directly upon the stone (except fig. 23). The preparations (exept fig. 23) are fixed in chromo-aceto-osmic acid, stained according to Heidenhain's hæmatoxylin method (except fig. 24) and some of them afterwards with Delafield's hæmatoxylin. All the illustrations are taken from ganglion cells of

Homarus vulgaris.

Fig. 23. (Magnified 620 diameters; Zeiss F, 1. Camera lucida.) A & B Ganglion cells (from a ventral ganglion) isolated in acetic acid (1), glycerin (1), and aqua dest. (5), and stained with picro-carmine. *a* Space surrounding the central bundle of primitive tubes (*b*) of the process, and in which no structure is visible. *n, n* Neuroglia nuclei situated outside the sheath of the cell and the process.

» 24. (Magnified 620 diameters; Zeiss F, 1.) Part of a section through two ganglion cells (A & B) and the neuroglia (*nu*) situated between them. A & B Parts of the protoplasm of the two ganglion cells, in which a spongioplasmic reticulation is distinctly visible. *cm, cm* Neuroglia-sheaths of the cells. *a* Transsected, spongioplasmic fibres penetrating, from the sheaths into the protoplasm of the ganglion cells. *b* Such a fibre seen in connection with the sheath. *sp* Similar fibre subdividing and forming a reticulation. *nu* Neuroglia. *n* Neuroglia-nuclei.

» 25. (Magnified 350 diameters; Zeiss CC, 4.) Section through a large ganglion cell (of an abdominal ganglion) and the root of its process; the lightly staining contents of which partly originates from bundles of primitive tubes. *s, s* Fibres penetrating from the neuoglia-sheath into the protoplasm of the cell. *a* Similar fibres transsected.

» 26. (Magnified 750 diameters; Zeiss F, 2.) Section through a ganglion cell and a part of its prosess, the contents of which arises equally from the whole protoplasm of the cell by a convergence af primitive tubes. *nu* Neuroglia, *cm* Sheath of a ganglion cell.

Fig. 27. (Magnified 620 diameters; Zeiss F, 1.) Section through a large ganglion cell and the root of its process, the lightly staining contets (*a*) of which arises from a union of bundles of primitive tubes. *b* Transsected spongioplasmic fibres, with which the spongioplasmic walls separating the primitive tubes, unite. *cm* Sheath of a ganglion cell.

» 28 & 29. (Magnified 620 diameter; Zeiss F. 1.) Parts of successive sections through the same ganglion cell, selected from a series of sections to show how the lightly staining contents (*p, p*) of the process penetrates into the protoplasm of the cels surrounded for some distance by deeply staining fibres (*s, s'*). *b* Such fibres transsected. *a* & *s"* Similar fibres penetrating into the protoplam of the cell. *cm* Sheath of the cell. *v, v* Transsected bundles or small masses of primitive tubes. *nv* Cavities in the sheath of the cell into which cavities the protoplasm of the cell partly extends.

Plate IV.

The illustrations are drawn under camera lucida from the microscope directly upon the stone. The preparations (exept fig. 41) are fixed and stained in the same way as in Plate III. All the illustrations are taken from ventral ganglia of

Homarus vulgaris.

Fig. 30—33. (Magnified 350 diameter; Zeiss CC, 4.) Successive sections through the process of a ganglion cell, selected from a series of transverse sections through an abdominal ganglion. In fig. 30 the process is seen quitting the cell in fig. 33 it is divided into two branches (*a* & *b*). *p, c* Contents of the process; near the cell (fig. 30) this contents is lightly stained, in a greater distance from the cell (fig. 31) it becomes more deeply stained until it, at a certain distance (fig. 32), becomes quite dark. *pr* Protoplasm of the ganglion cell. *sf* Spongioplasmic fibres, penetrating into the protoplasm of the cell and (*sf'*) into the contents of the process. *sf"* Similar spongioplasmic fibres occurring in the external layers of the contents of the process. *cm* Neuroglia-sheath of the ganglion cell. *is* Inner neuroglia-sheath enveloping the process. *nm* Concentric layers of neuroglia surrounding the process.

Fig. 34. (Magnified 350 diameters; Zeiss CC, 4). Transverse section through the process of a ganglion cell, exhibiting the concentric layers of neuroglia (*nm*) surrounding the contents of the process (*pc*). *br* Transsected side-branch, issuing from the process the process.

» 35 & 36. (Magnified 620 diameters; Zeiss F, 1.) Two transverse sections through the same process of a ganglion cell; selected from a series of sections. *pc* Contents of the process, in which the transsected primitive tubes are distinctly seen. *sf* Transsected spongioplasmic fibres ocurring in the external layers of the contents. *br, br'* Transsected branches issuing from the process. *nm* Concentric layers of neuroglia. *n* Neuroglia-nucleus. *b & c* Similar nuclei deeply stained.

» 37. (Magnified 350 diameters; Zeiss CC, 4.) Section through a ganglion cell. *a* External, lightly stained, layer of the protoplasm. *b* Mesial deeply stained part of the protoplasm. *sf* Spongioplasmic fibres. *sf'* & *sf"* Similar spongioplasmic fibres surounding the lightly stained contents of the process, in the prototoplasm of the cell. *br, br* Branches of the process.

» 38. (Magnified 620 diameters; Zeiss F, 1.) Section through three ganglion cells (A, B, C) from an abdominal ganglion. *v, v'* Transsected masses of primitive tubes, situated peripherically in the protoplasm of the cell. *v"* Similar masses situated on the sides of the bundle of primitive tubes forming the contents of the process. *sf* Spongioplasmic fibres surrounding this bundle of primitive tubes. *s* Septa or fibres, partly separating the masses of primitive tubes. *s'* Small septa and fibres penetrating these masses. *n* Neuroglia-uncleus. *a, a* Lightly stained parts of protoplasm situated round the bundle of primitive tubes forming the contents of the process. *b* External lightly staining layer of the protoplasm. *p* Process of the ganglion cell (C). *cm* Sheath of a ganglion cell.

» 39. (Magnified 620 diameters; Zeiss F, 1.) Lateral section through the same cell as is illustrated in fig. 38, A. In this section lightly staining masses of primitive tubes are almost solely transsected. *a* Deeply staining protoplasm. *b* Place where the section has passed just on the border between the deeply staining protoplasm and the lightly staining mass of primitive tubes. *ss* Spongioplasmic septa and fibres occurring in the lightly staining mass. *ss'* Similar septum with a little deeply staining protoplasm. *v, v'* Small lightly staining masses. *cm* Sheath of the cell. *nm* Neuroglia.

Fig. 40. (Magnified 350 diameters; Zeiss CC, 4.) Section through a group of smallish ganglion cells in a thoracal ganglion. *a* Transsected processes issuing from the cells. *n sh* Perineurium enveloping the ganglion. *nm* Neuroglia occurring inside the perineurium.

» 41. (Magnified 750 diameters; Zeiss F, 2.) Section of the nucleus of a ganglion cell stained with *borax-carmine*. The nucleus has two nucleoli (*a* & *b*). *c* Thickenings in the chromatine-structure. *n* Membrane of the nucleus.

» 42. (Magnified 750 diameters; Zeiss F, 2.) Sections of nuclei of ganglion cells.

Plate V.

All the illustrations of this plate are drawn under the camera lucida and from the microscope directly upon the stone. They are all taken from preparations fixed in chromo-aceto-osmic acid, and stained with Delafield's hæmatoxylin (and partly also with eosin or picric acid in turpentine).

Fig. 43. *Nereis virens.* (Magnified 1150 diameters, Zeiss. Hom. im. $1/_{18}$, 2.) Ganglion cells, from a transverse section of the ventral nerve-cord. 1—5 ganglion cells in three of which processes (*pr*) are transsected. 2 Part of a ganglion cell. *a* Transsected nerve-tubes of the dotted substance. *b* Lightly stained part of the protoplasm of the cell 1. *cm, cm* Sheaths of the ganglion cells. *nu* Neuroglia. *n m* Neuroglia-septum penetrating between the ganglion cells, and sometimes dividing them into groups. *lg* Large dark granules in the protoplasm.

» 44 & 45. *Nereis virens.* (Magnified 1150 diameters; Zeiss. Hom. im. $1/_{18}$, 2.) Ganglion cells, from horizontal sections of the ventral nerve-cord. *a* Mesial, deeply staining part of the protoplasm. *b* Peripheral, lightly staining layers of the protoplasm. *sf* Spongioplasmic fibres, penetrating into the protoplasm. *cm, cm'* Neuroglia-sheaths enveloping the ganglion cells. *nu* Neuroglia. *nm* Thick neuroglia-septum.

Fig. 46 & 47. *Lumbricus agricola.* (Magnified 2070 diameters; Zeiss. Hom. im. $^1/_{18}$, 4.) Sections of ganglion cells of the ventral nerve-cord. *a* Lightly staining parts of the protoplasm. *npr* Nervous processes. *br, br'* Side-branches, given off from the nervous process. *ppr* Protoplasmic processes. *cm* Neuroglia-sheath of the ganglion cells. *nu* Neuroglia. *nur* Neuroglia retuculatation. *n* Nouroglia-nucleus.

» 48. *Patella vulgata.* (Magnified 2070 diameters; Zeiss. Hom. im. $^1/_{18}$, 4.) Section of a ganglion cell of the pedal nerve-cord. *npr* Nervous process. *nur* Neuroglia-reticulation.

» 49 & 50. *Patella vulgata.* (Magnified 1150 diameters; Zeiss Hom. im. $^1/_{18}$, 2.) Sections of ganglion cells; taken from transverse sections of the pedal nerve-cord. *npr* Nervous processes. *npr'* Roots of nervous processes. *br* Branches given off from the nervous process. *ppr* Protoplasmic process terminating under the perineurium. *pe* Perineurium. *n, n'* Neuroglia-nuclei. *n"* Nucleus situated outside the perineurium and belonging to it. *nur* Neuroglia-reticulation.

Plate VI.

All the illustrations of this plate are drawn from the microscope directly upon the stone under the camera lucida.

Fig. 51—54. *Patella vulgata.* (Magnified 1350 diameters; Zeiss F, 4.) Isolated ganglion cells of the pedal nerve-cord. *npr* Nervous process. *gr* Yellow granules (containing hæmoglobin?) occurring in the protoplasm of the cells. *gr'* Simular granules exuded from the cells. *a* (fig. 52). Large granule of yellow substance, being formed on a place where such substance is exuded from the cell. *b* Side-branches, given off from the nervous processes. *ppr* Protoplasmic process. *a* (fig. 54) Short process or prolongation of the nucleus of the ganglion cell. (The cells were isolated in acet. ac. (1), glycerin (1), aqua dest. (5) or in weak solution of alcohol and stained with ammonia-carmine).

» 55—57. *Phallusia venosa.* (Magnified 1350 diameters; Zeiss F, 4) Isolated ganglion cells from the brain. Fig. 55 & 56 Peripheral cells. Fig. 57 Tripolar cell situated in the dotted substance. *npr* Nervous process. *a, a', b* Side-branches given of from the nervous processes. *ppr* Protoplasmic processes. (The cells were isolated in weak solution of alcohol or in potassium-bichromate (0.003 %)).

Fig. 58. *Corella parallelogramma.* (Magnified 1350 diameters; Zeiss F. 4.) Section of a ganglion cell situated peripherically in the brain. *n pr* Nervous process. *a* Lightly staining parts of the protoplasm. (Fixed in osmic acid (1 %) and stained with picro-carmine.)

» 59. *Homarus vulgaris.* (Magnified 350 diameters; Zeiss CC, 4.) Isolated ganglion cell of a ventral ganglion. *a & b* Lightly staining masses of primitive tubes. *c* Bundle of primitive tubes forming the contents of the nervous process (*npr*). *br* Side-branches given off from the process. (Macerated in acetic acid (1), glycerin (1), aqua dest. (5) and stained with ammonia-carmine.)

» 60. *Homarus vulgaris.* (Magnified 750 diameters; Zeiss F, 2.) Isolated ganglion cell from a ventral ganglion. *n* Neuroglia-nucleus situated in the sheath of the cell. *npr* Nervous proceses. (Macerated in acetic acid (1), glycerin (1), aqua dest. (5) and stained with picro-carmine.)

» 61. *Homarus vulgaris.* (Magnified 1350 diameters; Zeiss F, 4.) Part of a section through the dotted substance of a thoracal ganglion, exhibiting the origin of a bundle (*a*) of nerve-tubes or rather primitive tubes in the dotted substance. *a* Bundle of primitive tubes, longitudinally transsected. *b* Longitudinally transsected primitive tube. *c* Obliquely transsected primitive tube. *tpt* Transversally transsected primitive tubes. *n* Neuroglia-nucleus. (Fixed in chromo-aceto-osmic acid, stained with neutral borax-carmine.)

» 62. *Homarus vulgaris.* (Magnified 620 diameters; Zeiss F, 1.) Part of a section through the dotted substance of an abdominal ganglion, exhibiting the mixture of large and small tubes in the dotted substance. *ds* The real dotted substance consisting of very small tubes, which are transversally, or obliquely, or partly longitudinally transsected. *tnt, tnt'* Transversally transsected large nerve-tubes. *a & b* Bundle of nerve-tubes longitudinally transsected. *nt* Large nerve-tube. *nt'* Large nerve-tubes, subdividing dichotomously. *nu* Neuroglia. *n, n', n''* Neuroglia-nuclei. *x* Large nucleus.

Plate VII.

The illustrations of this plate are drawn from the microscope directly upon the stone, under the camera lucida.

Fig. 63 & 64. *Homarus vulgaris.* (Magnified 350 diameters, Zeiss CC, 4.) Parts of a transverse (fig. 63) and a longitudinal (fig. 64) section through ventral ganglia. $a—g$ Nervous processes from ganglion cells or branches springing from such processes. h Subdividing part of a nerve-tube. i Varicose nervous fibril. vf Varicose nervous fibril. 1—12 Parts of longitudinal nerve-tubes belonging to the longitudinal commissures. sf, sf' Slender nerve-tubes partly given off from these longitudinal large nerve-tubes. (From chromo-silver stained preparations.)

› 65. *Homarus vulgaris.* (Magnified 1150 diameters; Zeiss. Hom. im. $^1/_{18}$, 1.) Part of a transverse section through the dotted substance of an abdiminal ganglion. $a—f$ Nerve-tubes, some of which are seen to subdivide. (Fixed in chromo-aceto-osmic acid, stained in accordance with Heidenhain's hæmatoxylin method.)

› 66. *Nereis virens.* (Magnified 950 diameters; Zeiss. Hom. im. $^1/_{18}$, 1.) Part of a horizontal, longit. section through the dotted substance of the ventral nerve-cord. a The central vertical septum, dividing the nerve-cord longitudinally (compare fig. 10, a & fig. 14). tt, tt', tt'' Transverse, small and large, nerve-tubes crossing this septum. tt''' Transverse nerve-tubes passing to the root of a peripheral nerve. pt Large nerve-tubes passing to the root of the same nerve. t Small longitudinal nerve-tubes. lt Large longitudinal nerve-tubes. n, n' Neuroglia-nuclei, situated longitudinally (n) or transversally (n') in the dotted substance. (Fixed in chromo-aceto-osmic acid, stained with Delafield's hæmatoxylin.)

Plate VIII.

Fig. 67. *Nereis virens.* (Magnifid 620 diameters; Zeiss F, 1. Cam. luc. from the microscope direct upon the stone.) Part of a horizontal, longitudinal, section through the ventral part of the ventral nerve-cord. ds, ds The dotted substance on each side of the central mass of ganglion cells etc. *a* Neuroglia forming the central, vertical septum dividing the nerve-cord longitudinally (compare fig. 10, *a* & fig. 14). *lgc* Large ganglion cells with deeply staining protoplasm, with spongioplasmic fibres issuing from the sheaths of the cells, and with small lightly staining masses of primitive tubes (?). *np* Nervous process. *gc, gc'* Large ganglion cells with deeply staining protoplasm. *sgc* Small ganglion cells with lightly staining protoplasm. *sgc'* Ganglion cell of mesial size with a small mass of deeply staining protoplasm. *nf, nf* Neuroglia-fibres and longitudinally transsected tube-sheaths uniting with the sheaths of the ganglion cells. (Fixed and stained as fig. 66.)

Fig. 68. *Homarus vulgaris* (Magnified 100 diameters.) Ganglion cell from an abdominal ganglion; constructed from a series of transverse sections, partly by help of the cam. luc. (Fixed in chromo-aceto-osmic acid, stained in accordance with Heidenhains hæmotoxylin method.

» 69. *Homarus vulgaris.* (Magnified 132 diameters; Zeiss A A, 4. Cam. luc.) Ganglioncell from a thoracal ganglion, observed in a transverse section (chromo-silver staining); the proces could be traced directly into a peripheral nerve.

» 70. *Homarus vulgaris* (Magnified 350 diameters; Zeiss CC, 4. Cam. luc.) Unipolar ganglion cell from a chromo-silver stained section of a thoracal ganglion. The nervous process subdivides and is broken up into slender branches. The slender fibres issuing from the body of the cell are neuroglia-fibres issuing from its sheath.

» 71. *Lumbricus agricola* (Magnified 620 diameters; Zeiss F. 1. Cam. luc. from the microscope directly upon the stone.) Lateral part of a transverse section of the ventral nerve-cord. *a—h* Ganglion cells. *c* Ganglion cell containing two nuclei. *pe* Perineurium. *m* Muscles. *ct* Septa and fibres of comective tissue issuing from the perineurium. *nu* Neuroglia. *ds* Dotted substance. *nt, nt', nt"* Transsected nerve-tubes of various sizes. *lts, lts'* Sheaths of two of the three colossal nerve-tubes. *n, n'* Neuroglia-nuclei. (Fixed in chromo-aceto-osmic acid, stained with Delafield's hæmotoxylin).

» 72. *Lumbricus agricola.* (Magnified 750 diameters; Zeiss F, 2. Cam. luc.) Ganglion cell from a transverse section of the ventral nerve-cord.

Fig. 73. *Patella vulgata.* (Magnified 750 diameters; Zeiss F, 2.) Part of a transverse section through a pedal nerve-cord. *a—m, o* Ganglion cells. *n* Neuroglia-nuclei. *nn* Similar nuclei or cells from which fibres issue. *cn* Large ovoid or circular neuroglia-nuclei, more lightly staining than the smaller nuclei. *cn'* Similar nuclei situated close to the nervous processes of ganglion cells. *ds* Dotted substance. *p* Perineurium, enveloping the pedal nerve-cord. *nu* Neuroglia-reticulation extending inside this perineurium. *af* Fibres, neuroglia-fibres and protoplasmic processes, running from the ganglion cells *a* & *a'* towards the perineurium. *cct* Cells adhering externally to the perineurium. (Fixed in chromo-aceto-osmic acid, stained with Delafield's hæmotoxylin).

» 74 *Patella vulgata.* (Magnified 1150 diameters; Zeiss. Hom. im. $^1/_{18}$, 2.) Part of a transverse section through a pedal nerve-cord. *a—i* Ganglion cells. *ds* Dotted substance. *nu* Neuroglia reticulation. *n* Neuroglia-nuclei. *n'* Neuroglia-nuclei, adhering to the sheath of a ganglion cell. (Preparation the same as fig. 73).

» 75—78. *Patella vulgata.* (Magnified 750 diameters; Zeiss F, 2.) Ganglion cells with their processes. *n* Neuroglia nuclei. (Fig. 75 and 76 are isolated in glycerin (1), acetic acid (1), aqua dest. (5); fig. 77 and 78 are taken from in sections).

» 79—81. *Patella vulgata.* (Magnified 750 diameters; Zeiss F, 2.) Isolated neuroglia-cells.

Plate IX.

The illustrations of this plate (exept fig. 83—85) are drawn under the camera lucida, from the microscope directly upon the stone.

Fig 82. *Patella vulgata.* (Magnified 750 diameters; Zeiss F, 2.) Ganglion cells and their processes from a transverse section of a pedal nerve-cord. *gc* Ganglion cells. *a* Ganglion cell, the nervous process of which can be traced for some distance through the dotted substance. *n'* Neuroglia-nucleus adhering to the sheath of this process. *n* Neuroglia-nuclei. *nc* Larger less staining neuroglia-nuclei. (Fixed in chromo-aceto-osmic acid, stained with Delafield's hæmotoxyline. The section was stained with picric acid dissolved in turpentine to which was added a little abs. alcohol).

Fig. 83. *Patella vulgata.* (Magnified 570 diameters). Part of a transverse section through a pedal nerve-cord. *gc* Ganglion cell. *a* Ganglion cell. *n''* Neuroglia nuclei adhering to the sheath of this cell and its nervous pracess. *n'''* Three similar nuclei situated close together. *n* Neuroglia-nuclei. Quite similar nuclei (*n*) are also situated in the connective tissue outside the perineurium (*sh*). *n'* Larger, circular or ovoid, less staining nuclei situated in the same connective-tissue. *nc, nc'* Neuroglia-cells. *nc'''* Large, ovoid neuroglia-cell. *nf* Nerve-tubes or nervous processes. (Preparation the same as in fig. 82).

» 84. ***Lumbricus agricola.*** (Magnified 950 diameters; Zeiss. Hom. im $^1/_{18}$, 1.) Ganglion cells (*gc*) the nervous processes of which pass into the dotted substance (*ds*) and run longitudinally along the ventral nerve-cord. Taken from a horisontal longitudinal section of the ventral nerve-cord.

» 85. ***Patella vulgata.*** (Magnified 1150 diametars, Zeiss. Hom. im. $^1/_{18}$, 2.) Slender nerve-tubes and nervous fibrillæ, from a pedal nerve-cord, isolated in fresh state.

» 86. *Patella vulgata.* (Magnified 750 diameters, Zeiss F, 2.) Nerve-tubes and fibrillæ from a pedal nerve-cord, macerated in glycerin (1), acet. acid (1), aqua dest. (5). *a* & *b* Thick nerve-tubes giving off side-branches. *nc* Neuroglia cells.

» 87. *Corella parallelogramma.* (Magnified 1150 diameters; Zeiss. Hom. im. $^1/_{18}$, 2.) Part of a longitudinal section through the brain. *a—g* Ganglion cells *n* Neuroglia-nuclei. *n* Similar nucleus adhering to the sheath of a nervous process. (Fixed in osmic acid (1 %), stained with *picro-carmine*)

» 88. *Corella parallelogramma.* (Magnified 620 diameters; Zeiss F, 1.) Anterior part of a horizontal longitudinal section of the brain. *c* Ganglion cells. *gc* Ganglion cell situated in one of the anterior nerves. *n* Nuclei adhering to the inside of the perineurium. *n'* Neuroglia-nuclei situated in the sheaths of the nerve-tubes. (Fixed in *osmic acid* (1 %), stained with *picro-carmine*).

Fig. 89. *Phallusia venosa.* (Magnified 750 diameters; Zeiss F, 2.) Nerve-tubes and ganglion cells (*gc*) from a brain, macerated in weak solution of alcohol. *a* Nervous process from which side-branches are given off. *b—e* Nerve-tubes from which side-branches are given off.

» 90. *Amphioxus lanceolatus.* (Magnified 350 diameters; Zeiss CC, 4.) Transverse section of the spinal cord. *gc, gc'* Ganglion cells. *pp* Protoplasmic process. *vnt* Ventral colossal nerve-tube. *lnt* Lateral colossal nerve-tubes. *cg* Central groove (or canal). *f* Fibres crossing the grove. *nf* Fibres issuing from the epithelial cells surrounding the central groove and penetrating to the sheath, enveloping the spinal cord. *ns* Similar, thick fibres, united to strong bundles, one on each side of ventral, colossal nerve-tube. *p* Pigment. *nt* Slender nervetubes running transversally in the white substance. (Fixed in *chromo-aceto-osmic acid*, stained according to Heidenhain's hæmatoxylin method and afterwards with Delafields hæmatoxylin).

» 91. *Amphioxus lanceolatus.* (Magnified 750 diameters; Zeiss F, 2.) Transsected large nerve-tubes from the a transverse section of the spinal cord. (Preparation same as in fig. 90).

» 92. *Amphioxus lanceolatus.* (Magnified 620 diameters; Zeiss F, 1.) Ganglion cell (*gc*) from a transverse section of the spinal cord. *pp* Protoplasmic process penetrating to the sheath (*a*) enveloping the spinal cord. *pp'* Protoplasmic process, which possibly crosses the central groove, and which in this section is transsected. *cg* Bottom of the central groove (or canal). *e* Epithelium. *ee* Epithelial cell. *p* Pigment occurring at the bottom of the central groove. *ns* Fibres issuing from the epithelial cells and in which some pigment is deposited. (Fixed in *osmic acid* (1 %), stained with *ammonia-carmine*).

Plate X.

The illustrations of this plate are drawn under the camera lucida, from the microscope directly upon the stone. They are all taken from sections of the *spinal cord* of

Myxine glutinosa.

Fig. 93. (Magnified 132 diameters; Zeiss AA, 4.) Transverse section of the spinal cord. (To some extent composed from several sections of series of transverse sections, but each part is drawn under the cam. luc.) *gc* Ganglion cells. *lgc₁*, *lgc₂* Large ganglion cells. *mp* Mixed process (or nervous process?). *br* Branches given off from the mixed process. *pbr* Branches of the protoplasmic processes. *vnr₁*, *vnr₂*, *vnr₃*, *vnr₄* Ventral nerve-roots, which are not transsected in the same section, but which are drawn here to show their possition. *nbr* Side branch given off from a nerve-tube, of the ventral nerve-root. *nf* Nerve-tubes. *tnt* Transsected nerve-tubes. *v* Vacuoles which possibly also are transsected nerve-tubes. *Mnt* Müllers nerve-tubes transsected; those in the one half of the spinal cord are only illustrated. *nc* Neuroglia cell. *nn* Neuroglia nuclei, situated in the grey substance. f_1, f_2 Neuroglia fibres. *ep₁*, *ep₂*, *ep₃* Processes from the cells surrounding the central canal. *n* Nuclei of the sheath enveloping the nerve-tubes of the dorsal nerve-root.

Fig. 93. A. The transsected spinal ganglion, through which the nerve-tubes of the dorsal nerve-root pass. At *a* the nerve is somwhat shortened on account of the limited space of the plate. *sh* Wall of connective tissue surrounding the cavity in which the spinal cord is situated. *sgc* Transsected ganglion cells. *n"* Nuclei situated in the sheaths of the ganglion cells. *vr* Ventral ramus issuing from the ganglion. *dr* Dorsal ramus issuing from the ganglion. *n'* Nuclei situated in the sheaths of the nerve-tubes of the latter ramus. (Hardened in *Potassium bichromate* (2—3 %), stained with neutral *borax-carmine*).

» 94. (Magnified 154 diameters; Zeiss CC, 1.) Part of a transverse section through the spinal cord. *mp* Mixed process of a ganglion cell (which was not, hower, seen in the section). *pbr* Protoplasmic branches given off from this process. *gc₁*, *gc₂* Ganglion cells. *npr* Nervous processes. *ppr* Protoplasmic processes. *nf*, *nf₁*, *nf₂* Nerve-tubes or fibrillæ some of which (nf_1, nf_2) have varicose thickenings and give off branches. *cec* Cell of the epithelium surrounding the central canal. *grs* Border of the grey substance. (Stained according to the chromo silver method, vide p. 77).

» 95. (Magnified 132 diameters; Zeiss AA, 4.) Part of a transverse section of the spinal cord. The letters have the same signification as in fig. 93. *grs* Border of the grey substance. *eec* Epithelium surrounding the central canal. *npr* Nervous process. *nbr* Sidebranch of the nervous process. *mbr* Main branch of the mixed process from *lgc*. *br$_1$*, *br$_2$* Branches given off from the same process. *nt* Obliquely transsected nerve-tube from the ventral nerve-root (*vnr$_2$*). (Fixed in saturated aquous solution of *picric acid*, stained according to Heidenhain's hæmotoxylin method).

Fig. 96. (Magnified 950 diameters; Zeiss. Hom. im. $^1/_{18}$, 1). The ganglion cell *lgc* of fig. 95, more lighly magnified. *mp* Mixed process. *pp* Protoplasmic process. *v, v'* Transsected, lightly staining tubes (probably bundles of primitive tubes). *cm* Sheath enveloping the cell. *a* Cavity, filled with a lightly staining, reticular substance, between the sheath and the protoplam of the cell.

97—99. (Magnified 950 diameters; Zeiss. Hom. im $^1/_{18}$, 1.) Transsected ganglion cells of a spinal ganglion. *n, n'* Nuclei situated, generally, inside the sheath enveloping the cell. Nucleus situated in the sheath of a process. *gr* Dark granules, occurring inside the sheaths (possibly artifical products in the sections). *v* Transected, lightly staining tubes (probably bundles of primitive tubes). *x* Hyaline organ (or cell?) situated in the surface of the cell. *N* Elongated nucleus situated in this organ. *N'* Another nucleus (?) situated in the same organ. *pr* Process issuing from a cell; at *a* the sheath of this process is only transsected, we do not, therefore, see the origin of the process-contents in the protoplasm of the ganglion cell. *pn* Nucleus situated in the sheath of the process. *nt* Nerve-tube. *tn* Nucleus situated in the sheath of the nerve-tube. (Hardened in *potassium bicromate* (2—3 $^0/_0$) stained with *neutral borax-carmine*).

100. (Magnified 1150 diameters; Zeiss. Hom. im. $^1/_{18}$, 2.) Transsected nerve-tubes from a transverse section of the white substance. *lnt* Large nerve-tubes (Müller's nerve-tubes) *lnt'* Such tubes in which a slight concentration towards a deeply staining axis is visible. *nt* Transsected small nerve-tubes. *t* Tubes or cell-processes running transversally between the longitudinal nerve-tubes. *f* Neuroglia fibres, which are, however, slightly visible in this section. (Fixed in saturated aqvous solution of *picric acid*, stained according to Heidenhain's hæmatoxylin method).

14*

Fig. 101. (Magnified 620 diameters; Zeiss. F, 1.) Part of a transverse section through the white substance (of the ventral side of the spinal cord. The neuroglia fibres (f) running transversally between the transsected nerve-tubes are here very distinctly visible. a External layer where these unite with the sheath (pm), enveloping the spinal cord. lnt Transsected large nerve tubes, where the contents (c) has skrunk into one side and is deeple stained. c' Contents (skrunk in the same way) of the small longitudinal nerve-tubes. nf Slender nerve-tubes running in the external layers of the white substance. (Hardened in alcohol, stained according to Heidenhains's hæmotoxylin method).

Plate XI.

The illustrations of this plate (exept fig. 109 & 113) are taken from *chromo-silver* stained sections of the spinal cord of *Myxine glutinosa*. They are drawn under the camera lucida.

Fig. 102. (Magnified 154 diameters; Zeiss CC, 1.) Part of a transverse section through the spinal cord of *Myxine*. gc Ganglion cell. mpr Mixed process. br_1, br_2, br_3, br_4 Branches given off from this process. pr Protoplasmic (?) process, which, however, seem to return into the white substance. mpr_2 Mixed process of a cell which is not seen in the section. ppr' Protoplasmic process of a cell which is not seen in the section. nf, nf'', nf''' Nerve-tubes or fibrillæ, running transversally, in various directions in the grey and white substance. vnr Ventral nerve-root. v Vacuole, possibly transsected nerve-tube. pec Epithelial cell with process, stained for considerable distance through the grey tubstance. grs Border of the grey substance.

» 103. (Magnified 150 diameters; Zeiss CC, 1.) Part of a transverse section through the spinal cord of *Myxine*. gc_1, gc_2, gc_3 Ganglion cells. np Nervous processes. ppr Protoplasmic processes. a Place where a branch is given off from the nervous process np'. nuc_1, nuc_2 Neuroglia cells. nf, nf', nf'' Nerve-tubes. ec Epethelial cells surrounding the central canal. grs Border of the grey substance. dnr Dorsal nerve-root.

» 104. (Magnified 200 diameters). Ganglion cell, from a transverse section of the spinal cord of *Myxine*. sh Sheath

Fig 105. enveloping the spinal cord. *grs* Border of the grey substance. *np* Nervous process. *ppr* Protoplasmic process. (Magnified 130 diameters). Ganglion cell, from a transverse section of the spinal cord of Myxine. *np* Nervous process or mixed process. *ppr* Protoplasmic process. *vnr* Nerve-tube of the ventral nerve-root. *sh* sheath enveloping the spinal cord.

» 106. (Magnified 130 diameters). Ganglion cell, from at transverse section of the spinal cord of *Myxine*. *vnr* Nerve-tube (of the ventral nerve-root) springing directly from the ganglion cell. *sh* Sheath enveloping the spinal cord.

» 107. (Magnified 130 diameters). Mesial part of a transverse section through the spinal cord of *Myxine*. Plenty of nerve-tubes (*nt*) are seen in the section crossing the transverse commissures. *cc* Central canal. *v* Vacuoles. *cee* Central epithelial cells with stained processes. *sh* Sheath enveloping the spinal cord.

» 108. (Magnified 200 diameters). Neuroglia cells with their processes from a transverse section of the spinal cord. The black body in the centre is probably produced artificially by the staining of several cells situated close together. *sh* Sheath enveloping the spinal cord.

» 109. (Magnified 500 diameters). Part of a transverse section through the white substance of the spinal cord, showing how neuroglia fibres gather round the nerve-tubes of the ventral nerve-roots. *p* Ventral periphery of the spinal cord. *grs* Border of the grey substance. *vnr* Nerve-tube of the ventral nerve-root. *br* Side-branch given off from this nerve-tube. *f* Neuroglia-fibres. *nc* Neuroglia-cells. *gc* Ganglion cell. *ppr* Protoplasmic process. (Fixed in saturated aqveous solution of picric acid, stained in accordance with Heidenhain's hæmatoxylin method).

» 110. (Magnified 500 diameters). Neuroglia-fibres running transversally between the large nerve-tubes (Müller's nerve-tubes) in the white substance; from a chromo-silver stained transverse section of the spinal cord of *Myxine*. *nuf* Neuroglia-fibres. *Mnt* Müller's nerve-tubes. *pe* Ventral periphery of the white substance.

» 111. (Magnified 80 diameters.) Part of a chromo-silver stained horisontal section of the spinal cord of *Myxine;* exhibiting the dichotomous subdivisions of the nerve-tubes of the nerve-root. *dnr* Dorsal nerve-root. *a* Nerve-tube subdividing outside the spinal cord. *pe* Periphery of the spinal cord. *cc* Central canal. *lnt* Longitudinal nerve-tubes. *lnt* Longitudinal nerve-tubes giving off side-branches. *snt* Subdividing nerve-tube. *snt'* Subdividing nerve-tube crossing the central canal.

Fig. 112. (Magnified 160 diameters.) Part of a chromo-silver stained horisontal section of the spinal cord. The letters have the same signification as in fig. 111. *br* Slender branches given off from the nerve-tubes of the dorsal nerve-root.

» 113. *Diagrame of the reflex-curve* The large arrows indicate the way the irritation of a sensitive nerve-tube has to pass to produce a reflex-movement. *SN* Centripetal (sensitive) nerve-tube. *dd* Dotted substance or interlacing of nervous fibrillæ in the central nerve-system. *MN* Centrifugal (motoric) nerve-tube.

The small arrows indicate the way small parts of the irritation of the centripetal (sensitive) nerve-tube pass to arrive in other parts of the central nerve-system. 1 Nerve-tube passing to the brain. 2 Longitudinal nerve-tube running along the spinal cord, whilst giving off side-branches. *snc* The nutritive centre of the centripetal nerve-tube (i. e. spinal ganglion cell). *cnc* The nutritive centre of a part of the fibrillæ forming the dotted substance or interlacing of nervous fibrillæ (i. e. ganglion cell of the central nerve-system). *mnc* The nutritive centre of the centrifugal nerve-tube (i. e. ganglion cell of the central nerve-system). *ppr* Nutritive (i. e. protoplasmic) processes sometimes issuing from the nutritive centres, and penetrating towards the periphery of the central nerve-system or towards blood-vessels to absorb nutrition. *a & b* Periphery of the central nerve-system.

Contents.

Introduction: Pag.

1. History 29
    - a) The structure of the nerve-tubes. 30
    - b) The structure of the ganglion cells. 32
    - c) The structure of Leydig's dotted substance 38
    - d) The combination of the ganglion cells with each other. 64
    - Dr. B. Rawitz's paper on the central nerve-system of the *Acephala*. . . 65
2. The material examined 72
3. Methods of investigation. 73

Description of my investigations:

1. The structure of the nerve-tubes in invertebrates 81
    - Summary 97
2. The structure of the ganglion cells, and their processes, in invertebrates . 98
    - Summary 120
3. The structure of Leydig's dotted substae. 122
    - Summary 144
4. The combination of the ganglion cells with each other, and the function of the protoplasmic processes 145
5. The nervous elements of *Amphioxus* and *Myxine* 149
6. The combination of the nerve-tubes with each other 164
7. The function of the ganglion cells 167

List of the principal Literature 172

Explanation of the plates 194

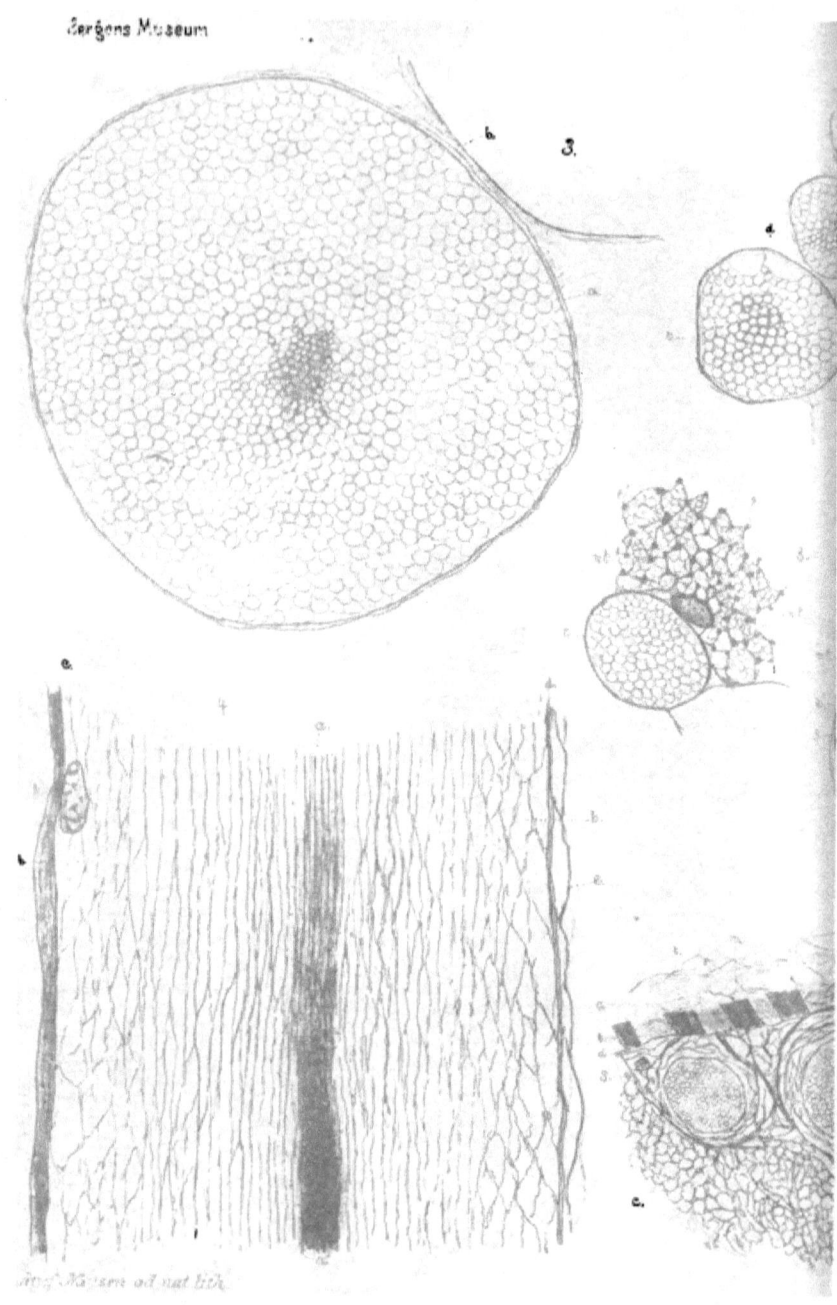

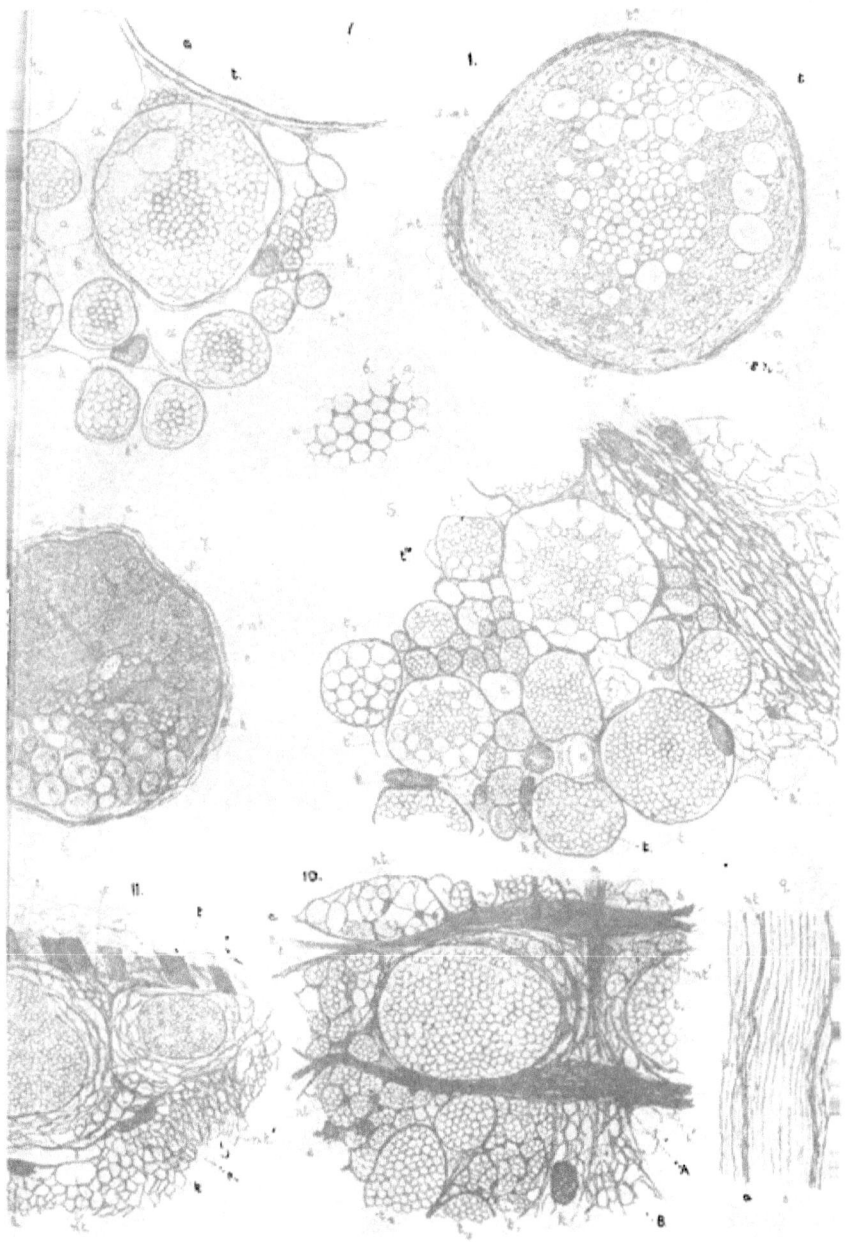

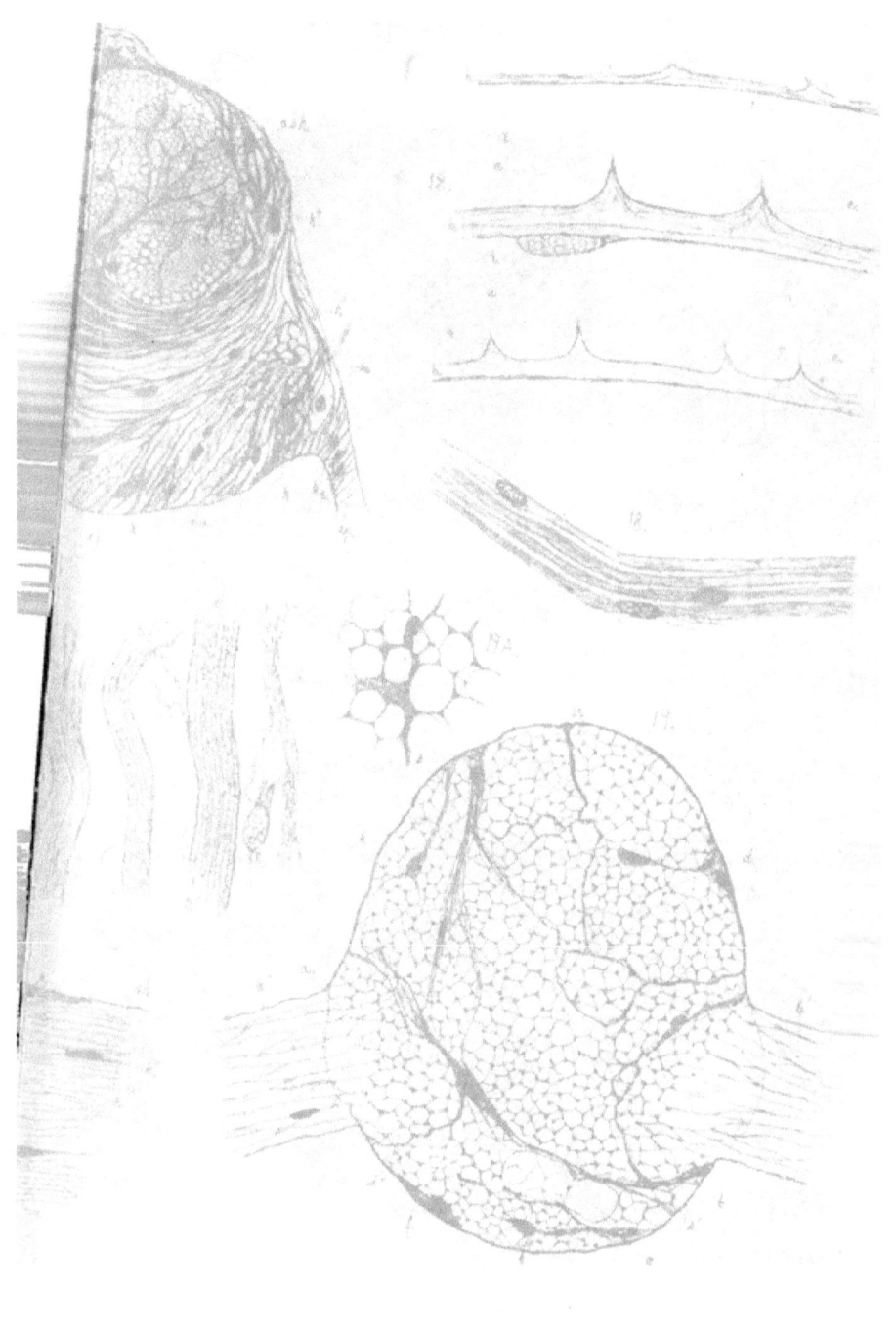

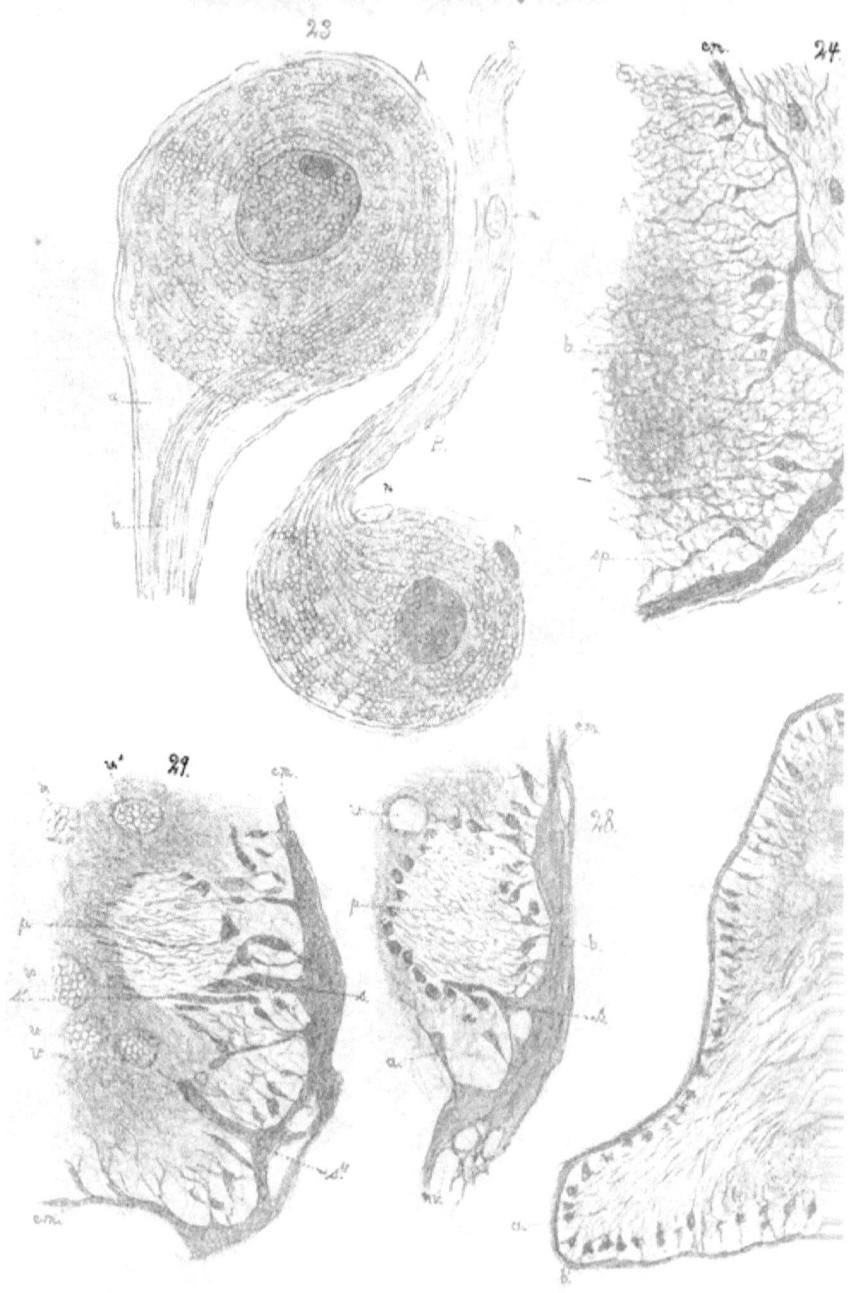

Fridtjof Nansen ad nat. lith.

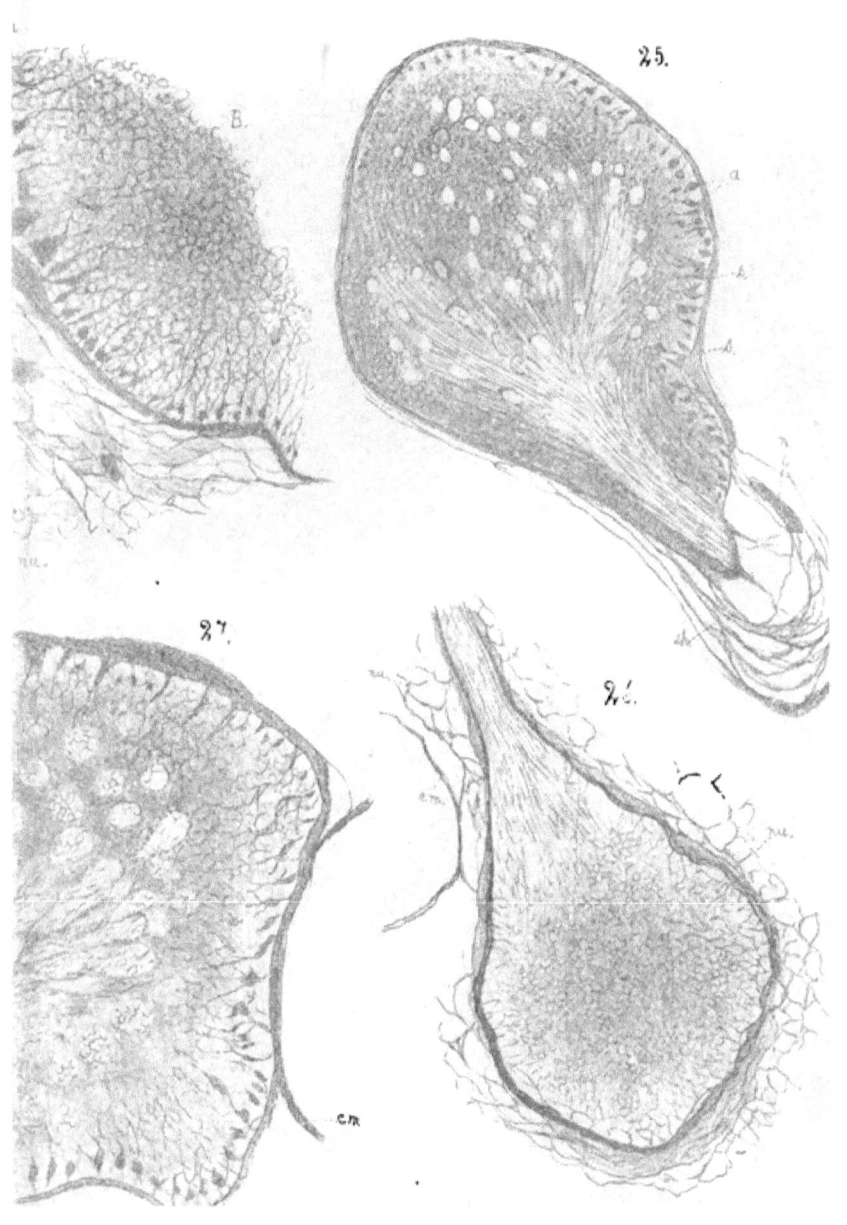

Pl. III.

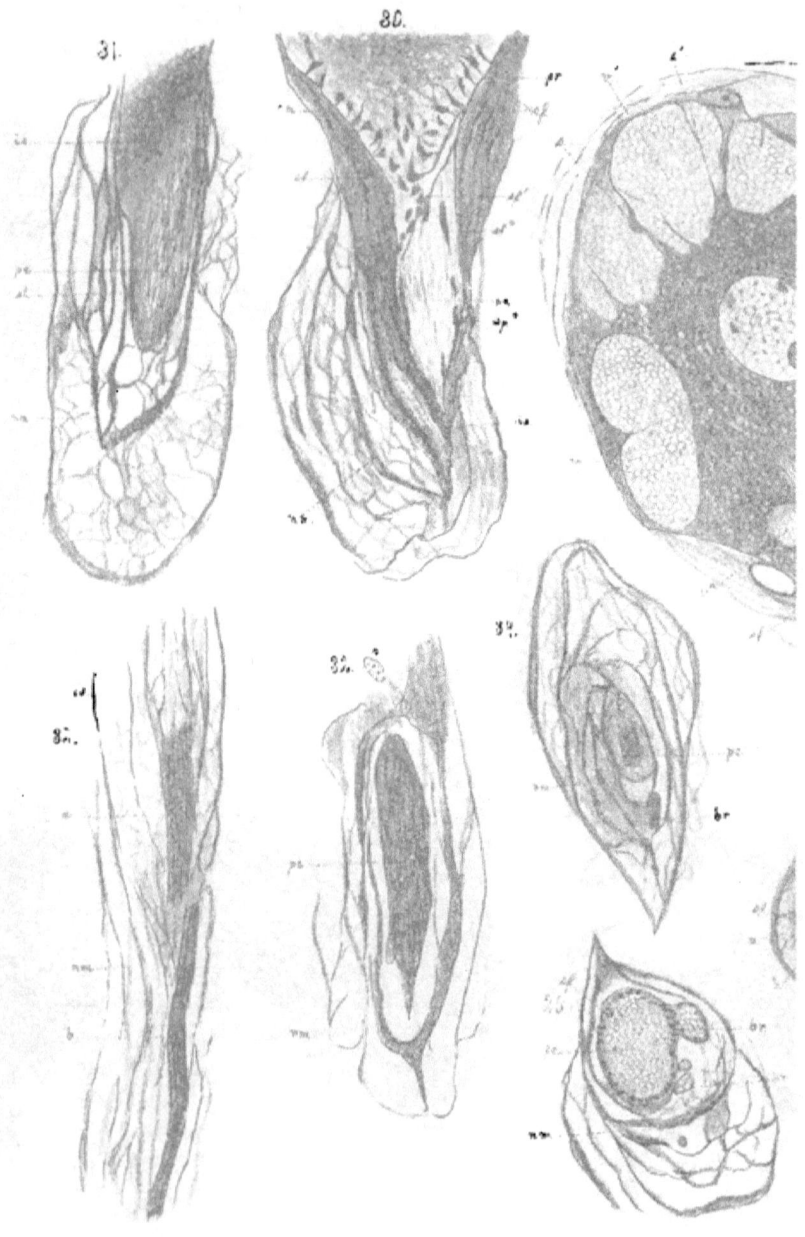

Fridtjof Nansen ad nat. lith.

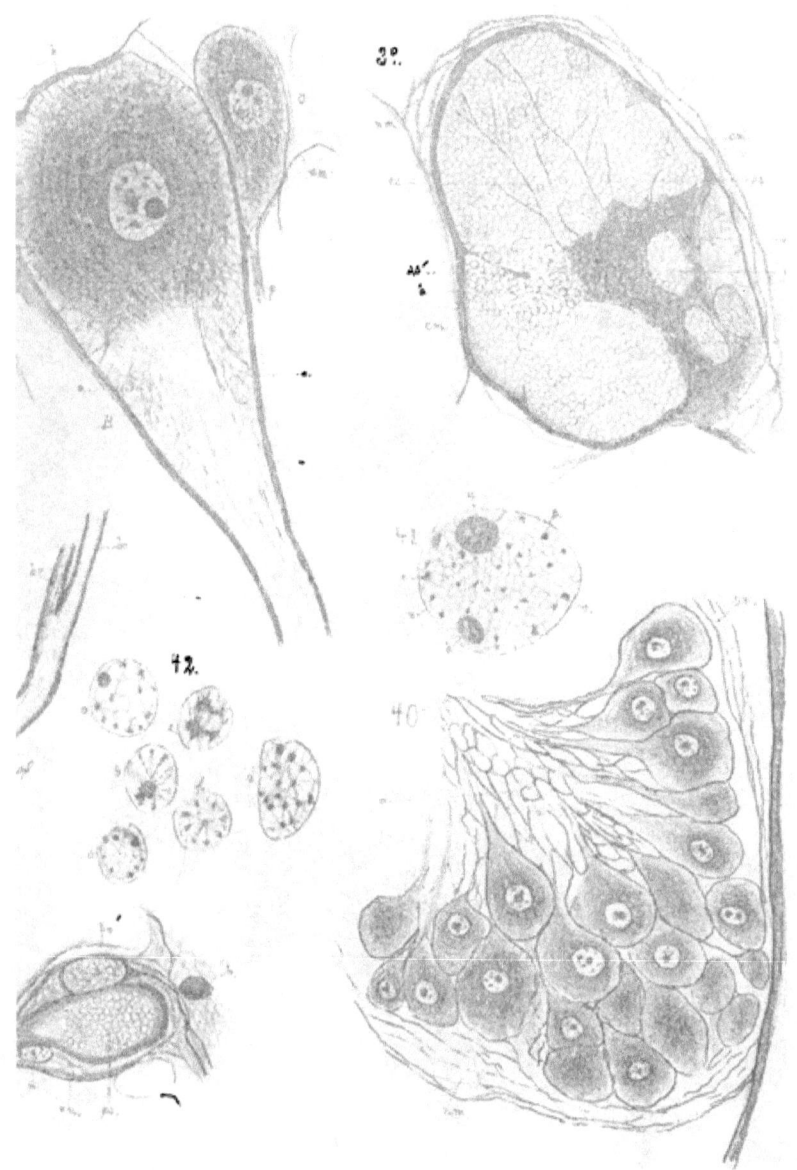

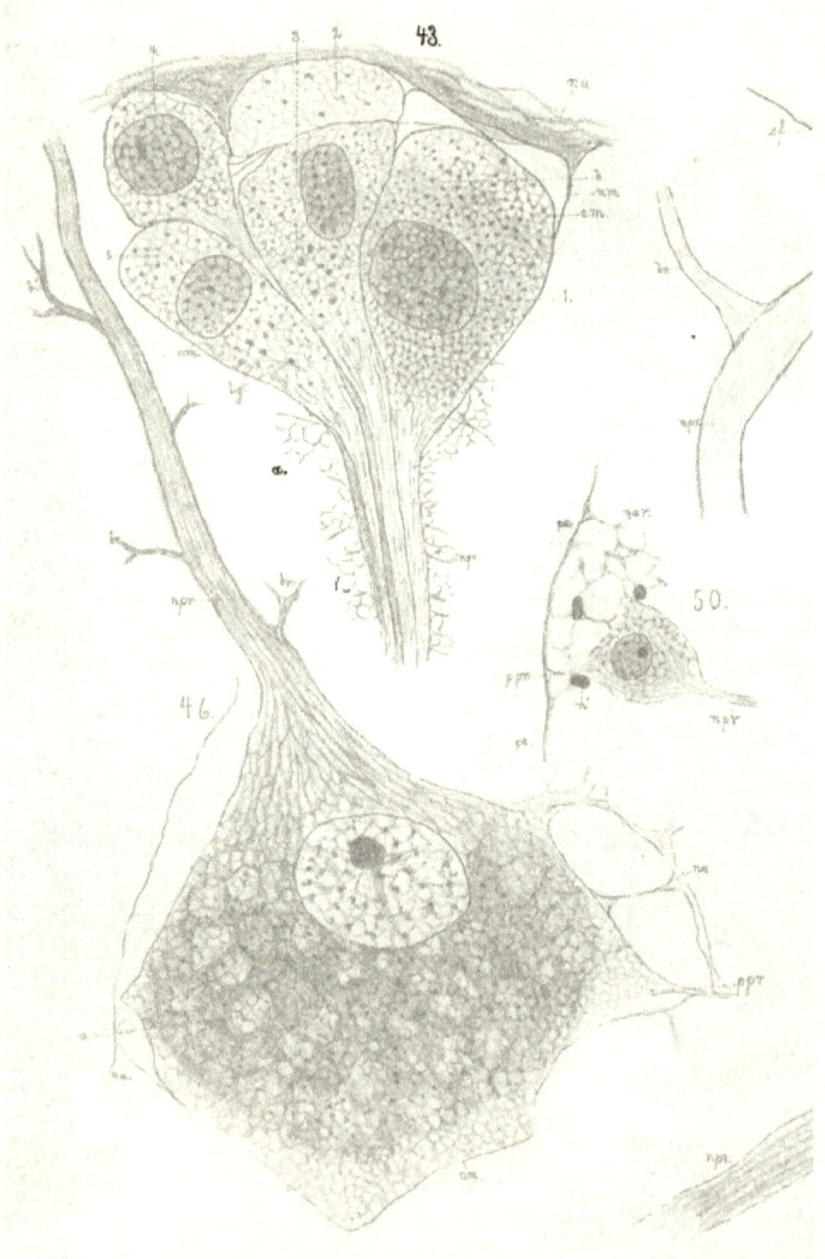

Bergens Museum.

Fridtjof Nansen ad nat. lith.

Pl. V.

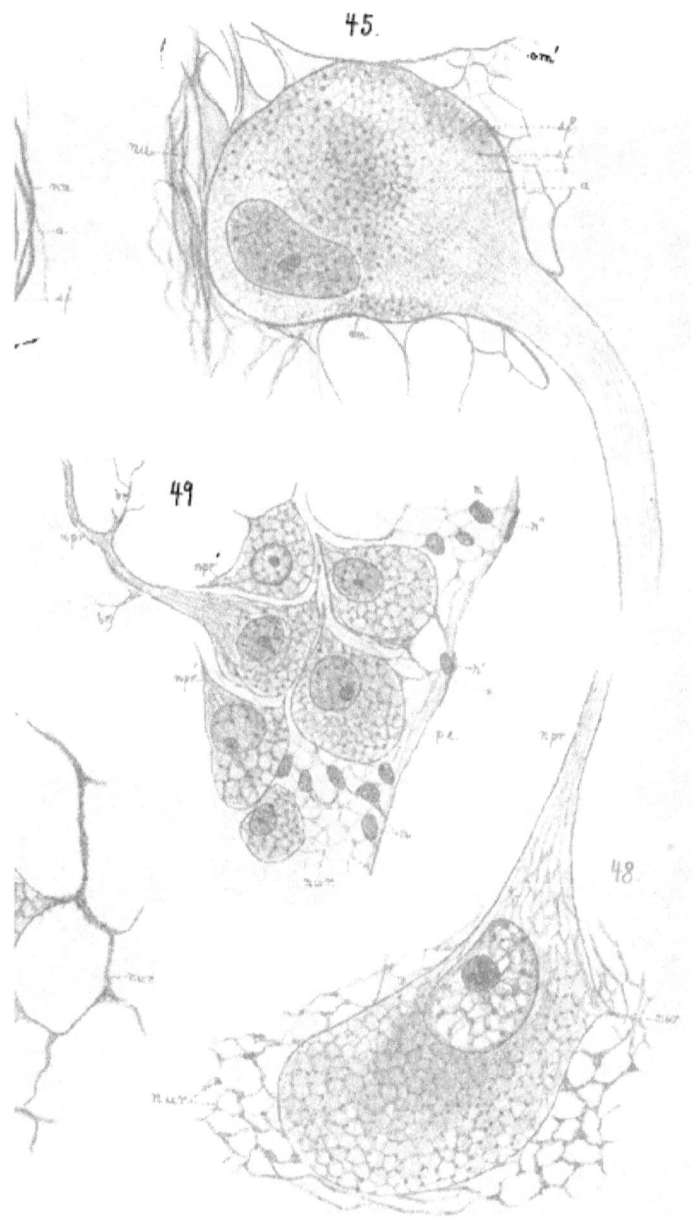

Bergens Museum.

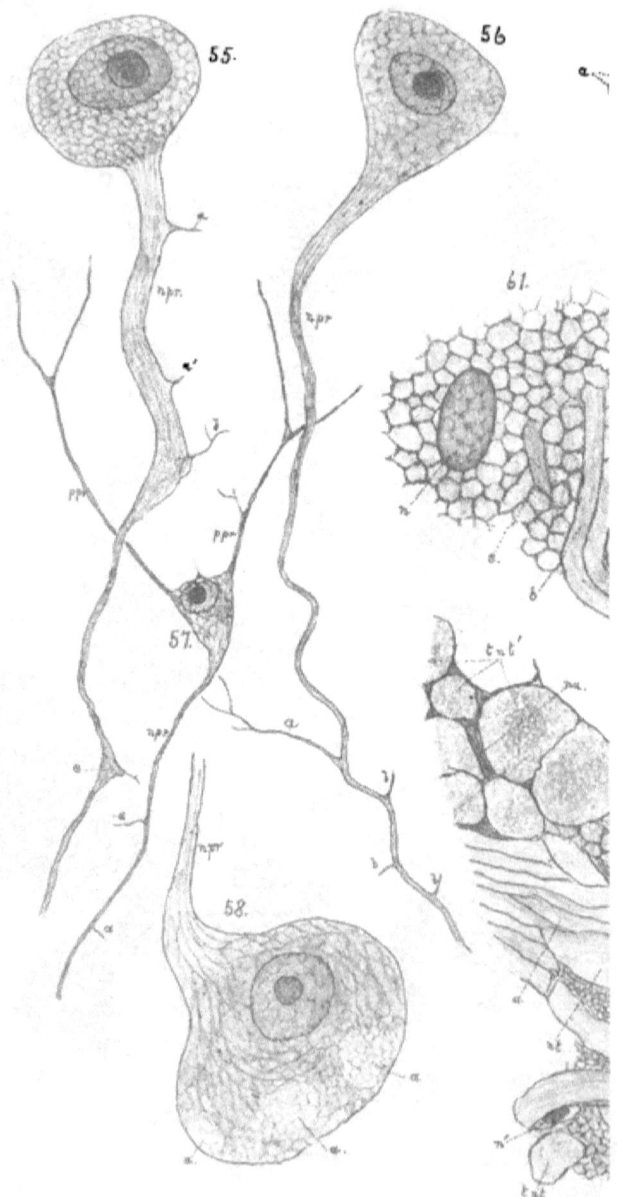

Fridtjof Nansen ad nat. lith.

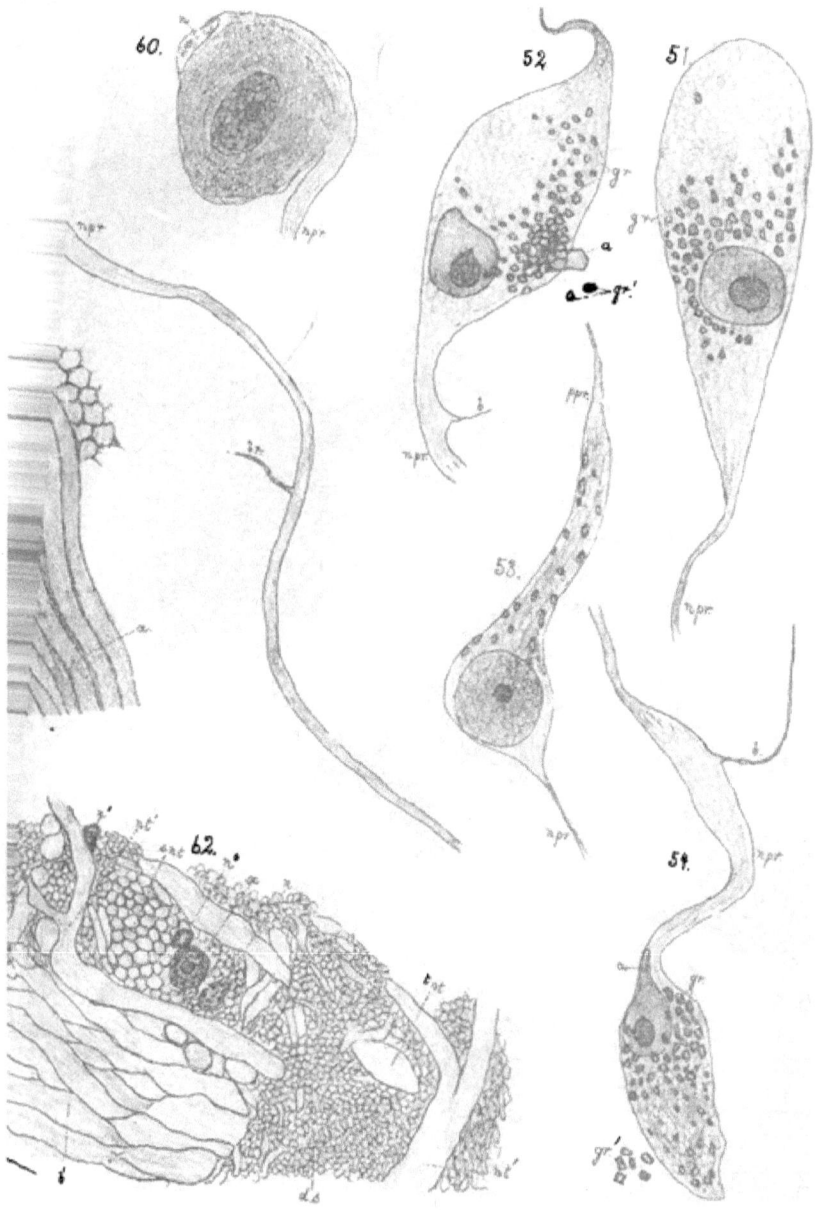

1

Bergens Museum

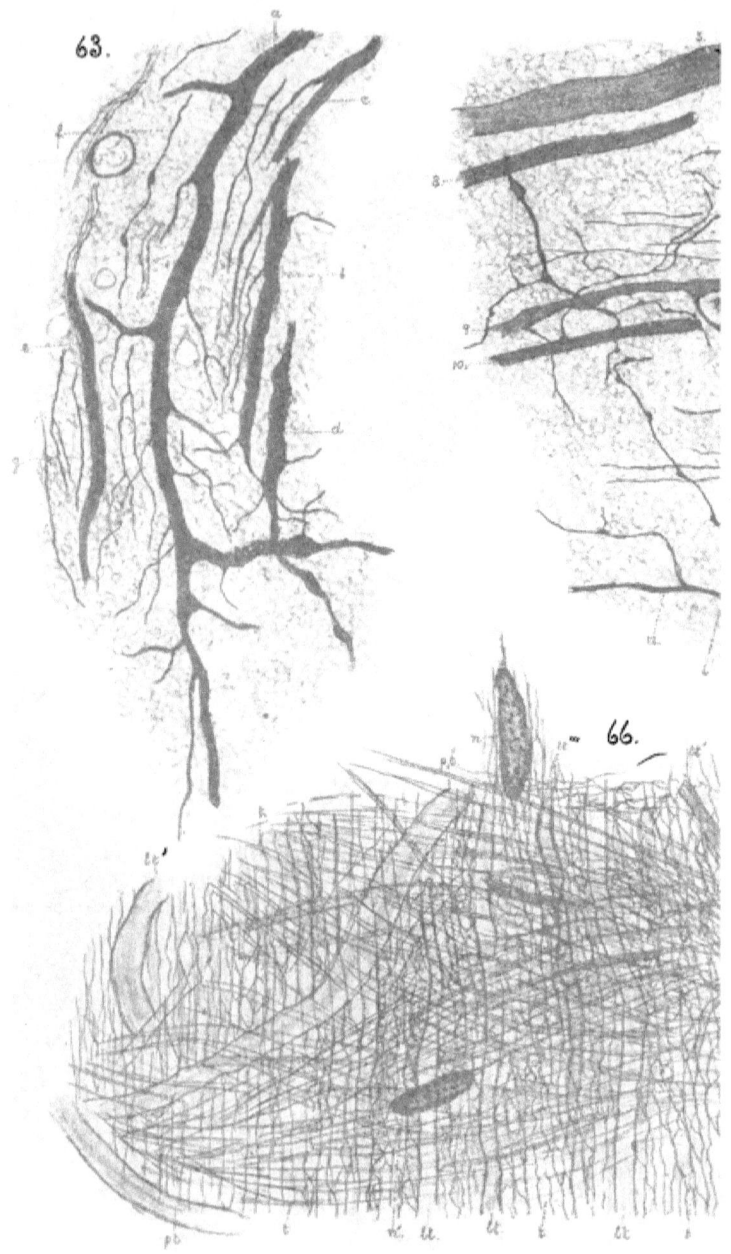

Fridtjof Nansen ad nat lith.

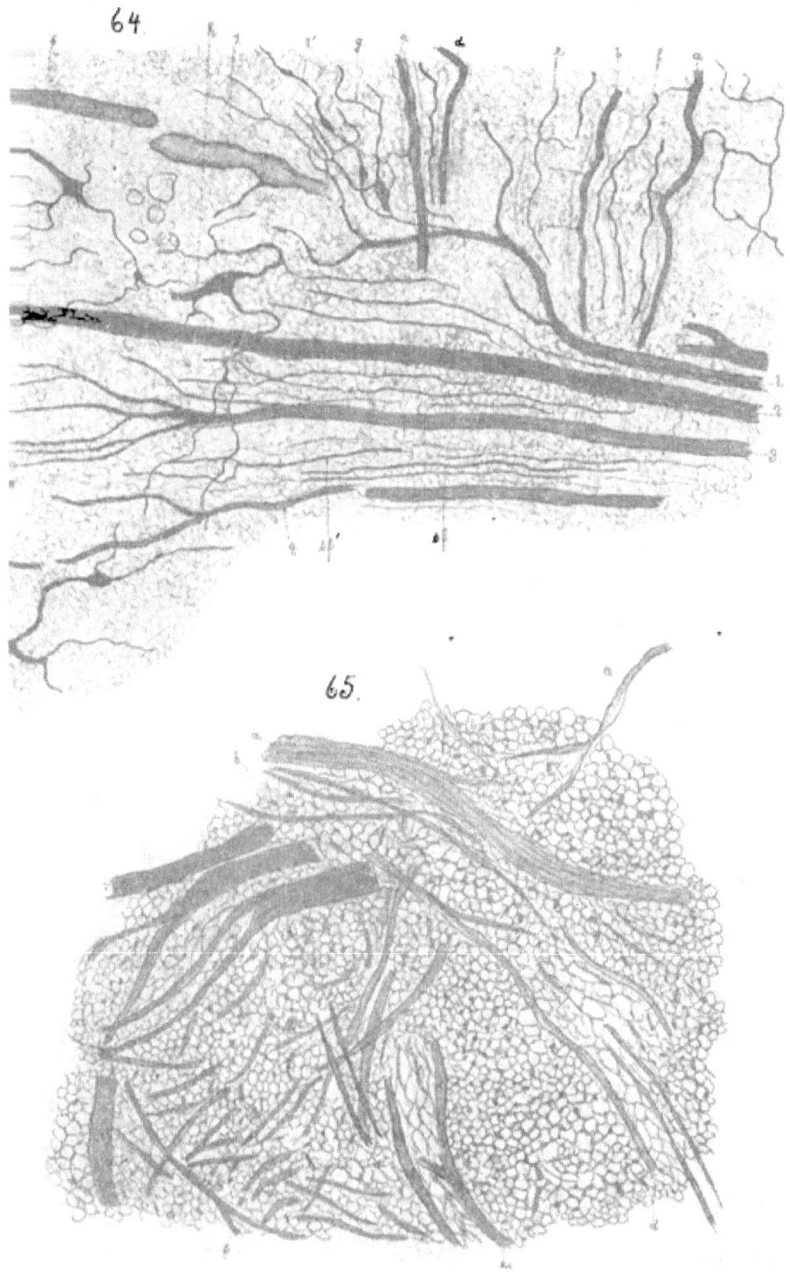

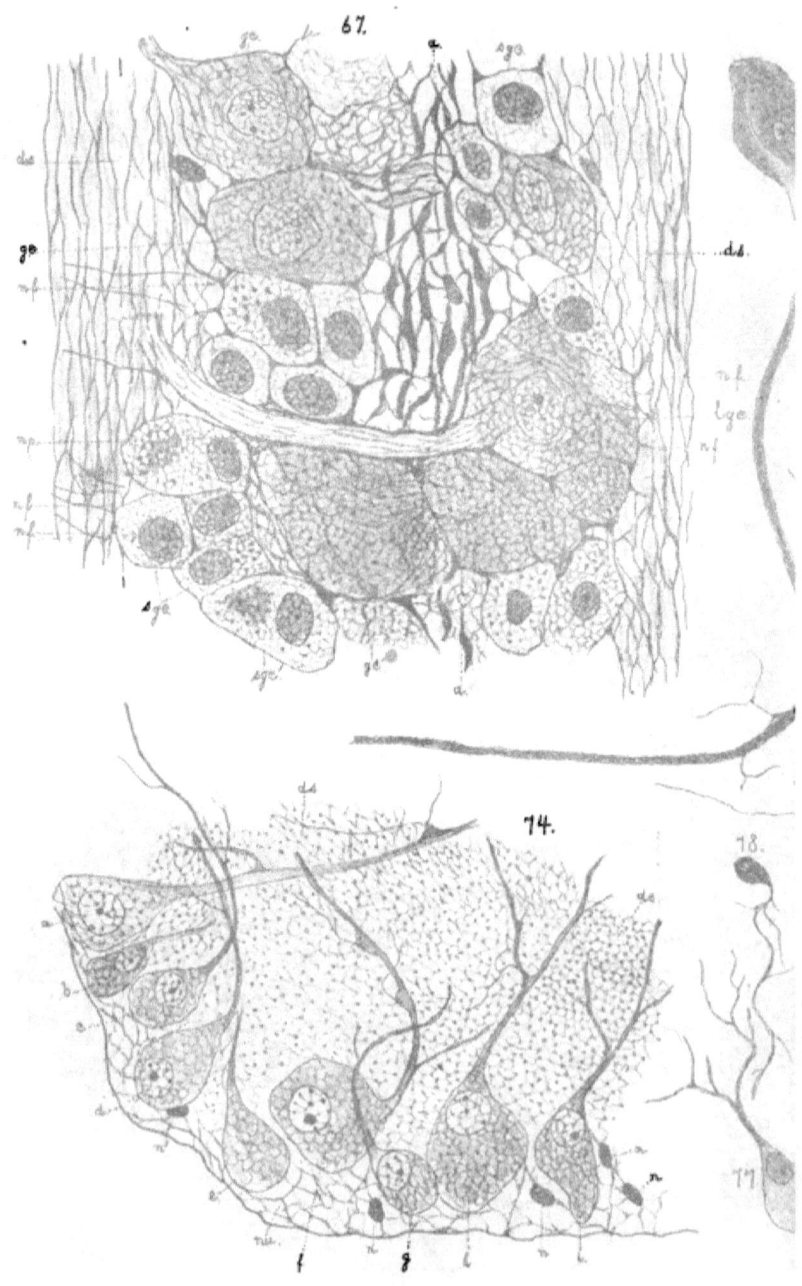

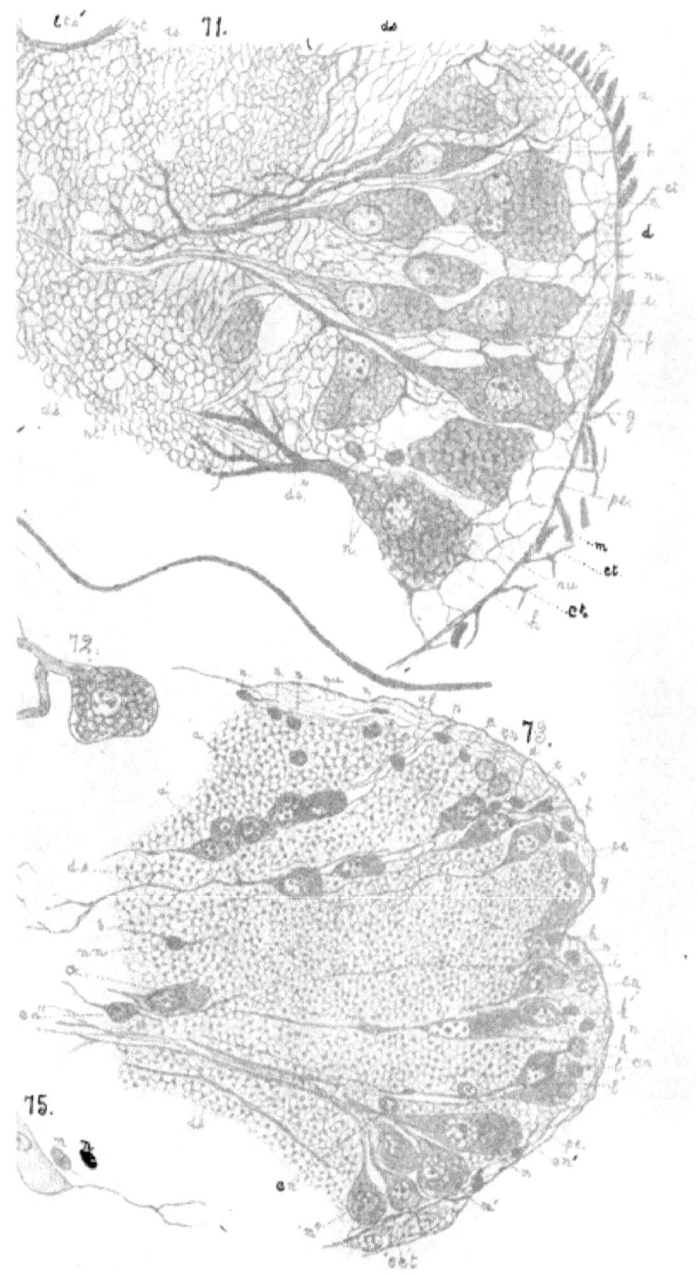

Pl. VIII

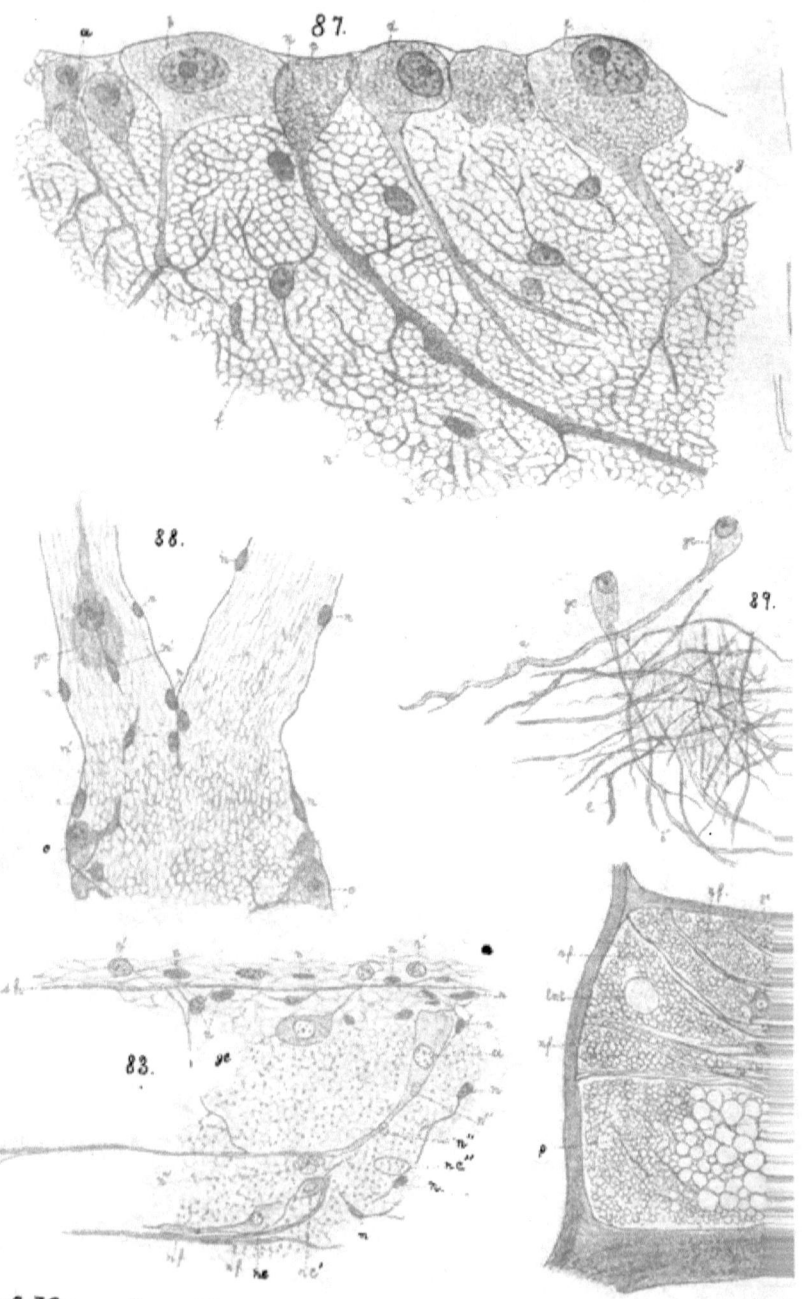

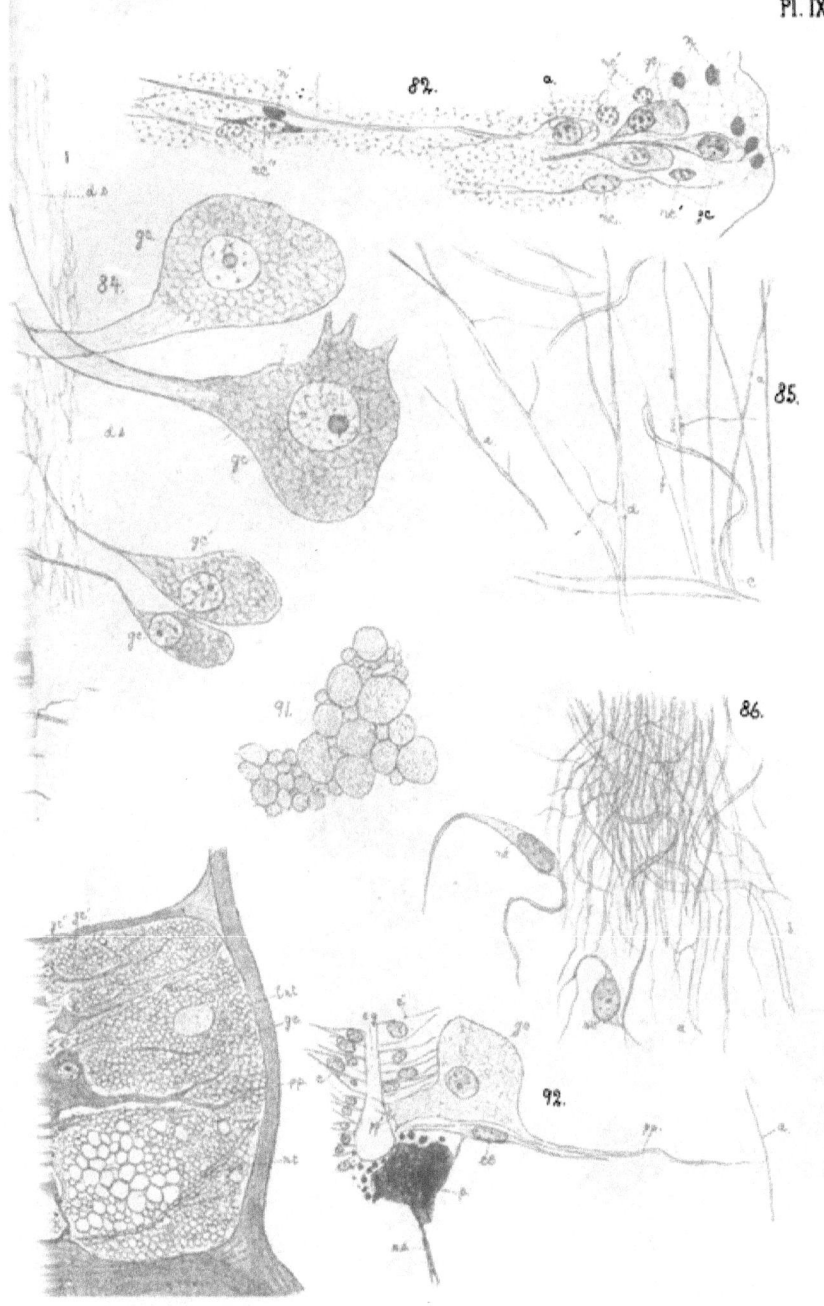

Pl. IX.

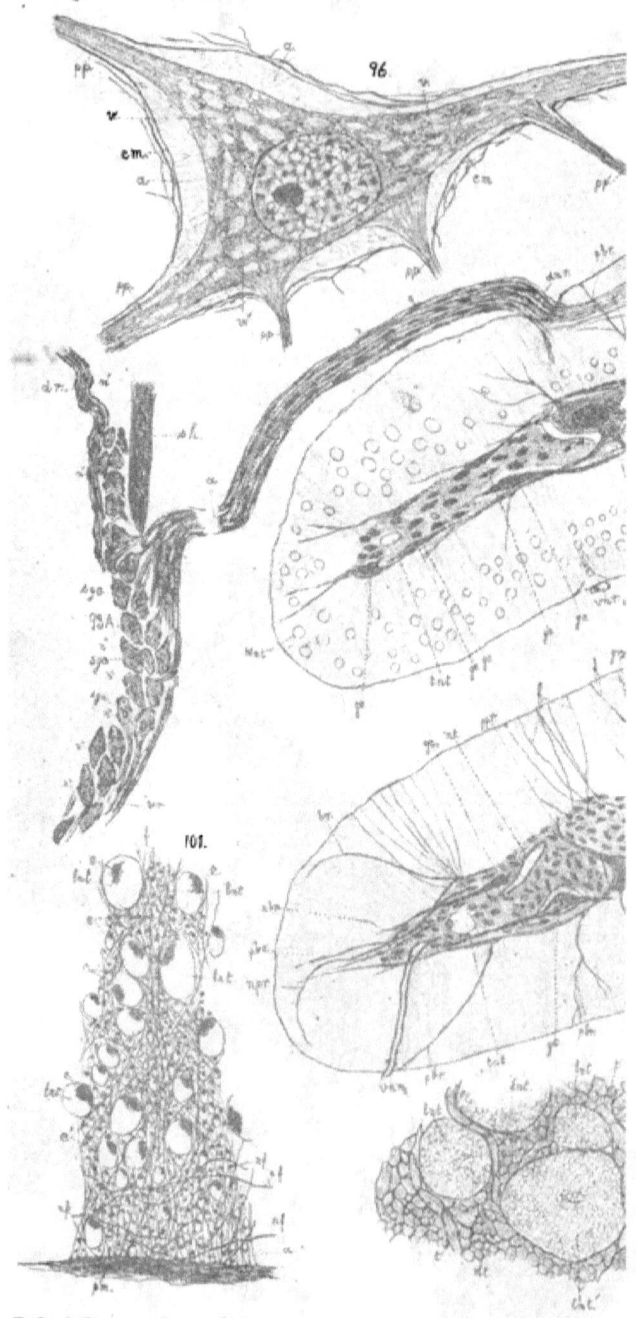

Fridtjof Nansen ad nat. lith.

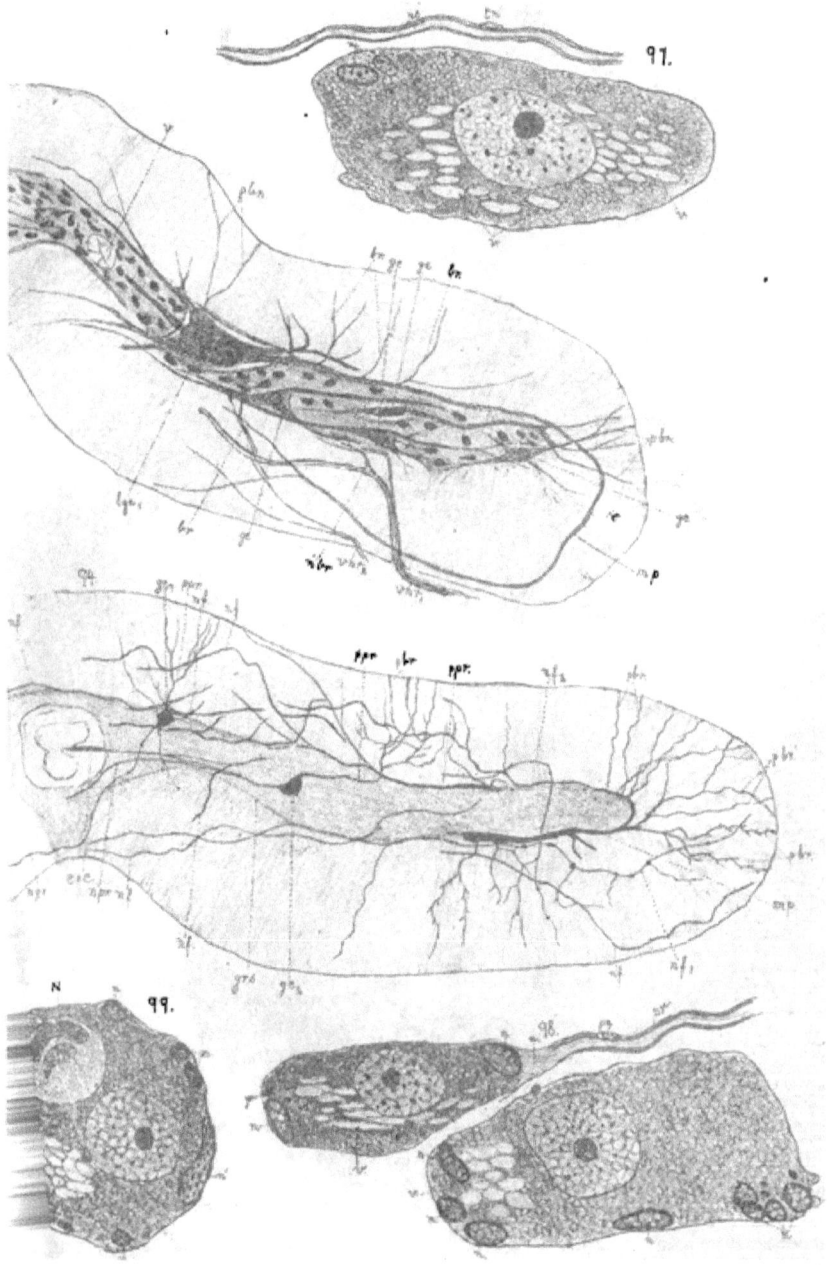

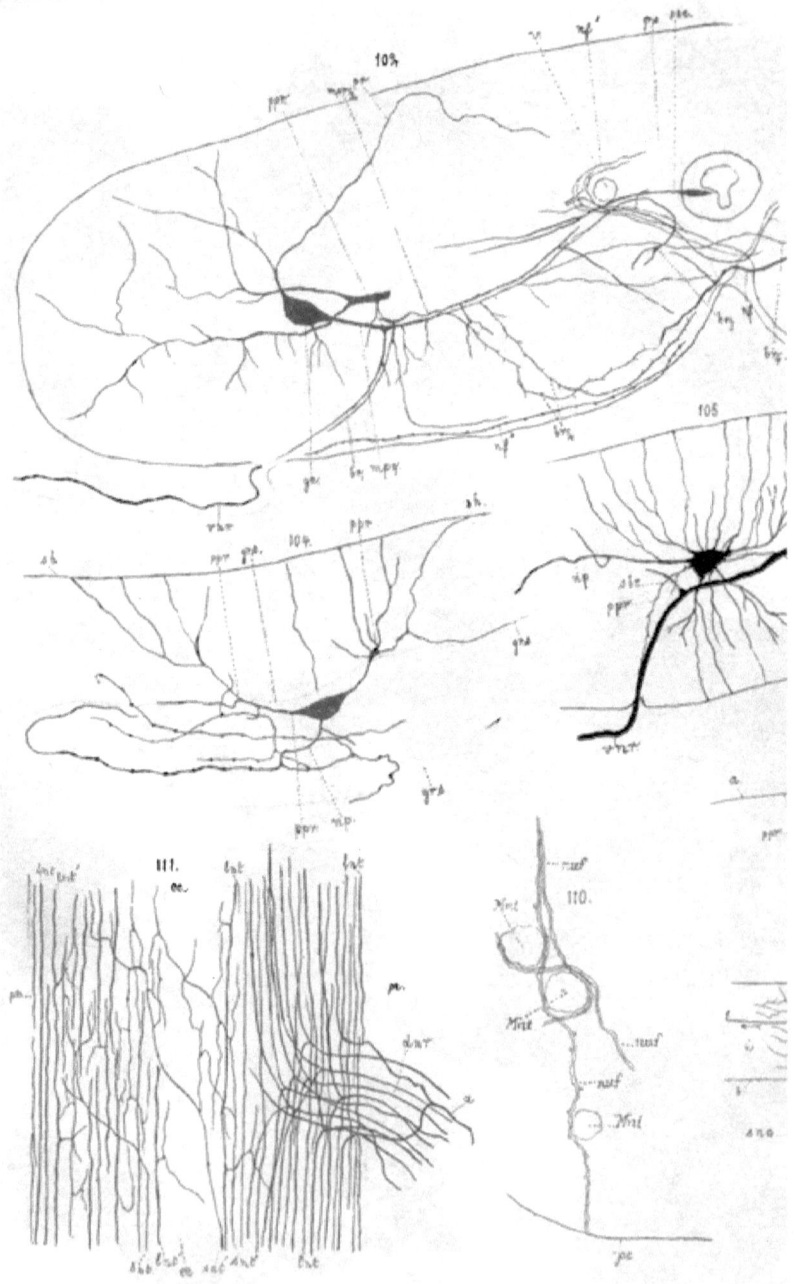

Bergens Museum.

Fridtjof Nansen ad nat. lith.

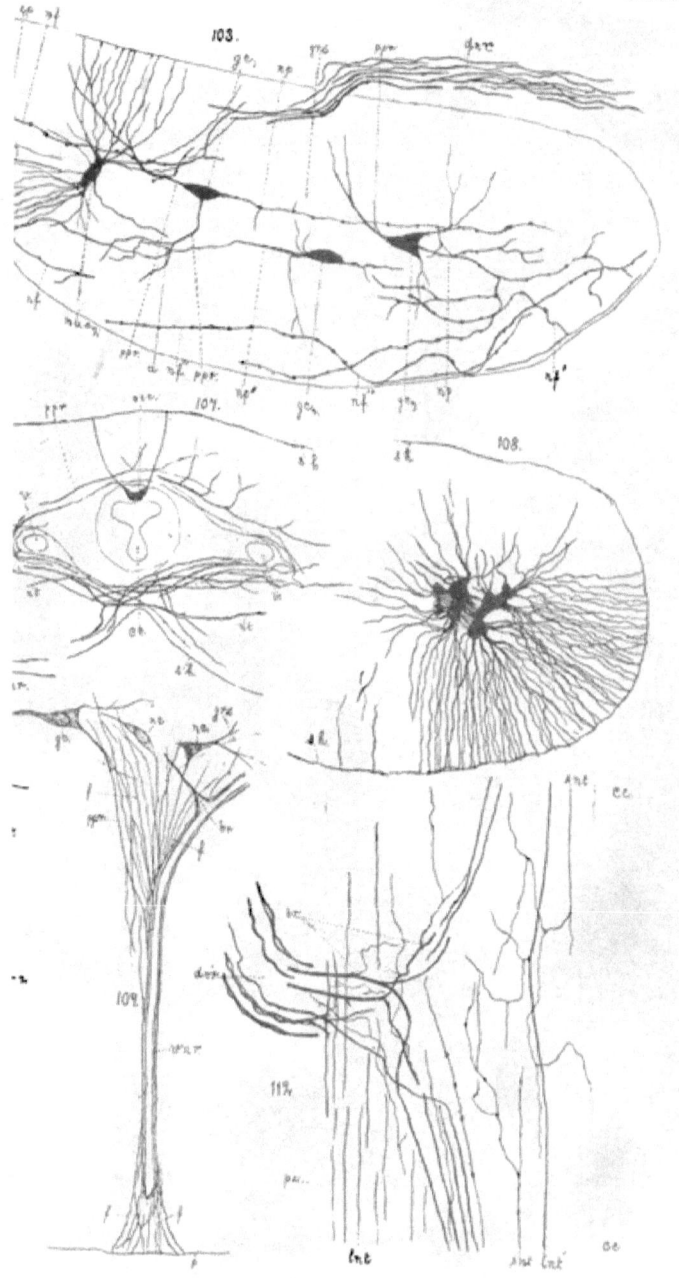

www.ingramcontent.com/pod-product-compliance
Lightning Source LLC
Chambersburg PA
CBHW021839230426
43669CB00008B/1014